Understanding the NEC3 ECC Contract

As usage of the NEC (formerly the New Engineering Contract) family of contracts continues to grow worldwide, so does the importance of understanding its clauses and nuances to everyone working in the built environment. Currently in its third edition, this set of contracts is different to others in concept as well as format, so users may well find themselves needing a helping hand along the way.

Understanding the NEC3 ECC Contract uses plain English to lead the reader through the NEC3 Engineering and Construction Contract's key features, including:

- Main and Secondary Options;
- the use of early warnings;
- programme provisions;
- payment;
- compensation events;
- preparing and assessing tenders.

Common problems experienced when using the Engineering and Construction Contract are signalled to the reader throughout, and the correct way of reading each clause explained. The way the contract affects procurement processes, dispute resolution, project management and risk management are all addressed in order to direct the user to best practice.

Written for construction professionals, by a practising international construction contract consultant, this handbook is the most straightforward, balanced and practical guide to the NEC3 ECC available. It is an ideal companion for Employers, Contractors, Project Managers, Supervisors, Engineers, Architects, Quantity Surveyors, Subcontractors and anyone else interested in working successfully with the NEC3 ECC.

Kelvin Hughes spent 18 years in commerc tors, then the past 19 years as a consultant, ship at the University of Glamorgan. He has since 1996, was Secretary of the NEC Users 1,200 NEC training courses.

D1355086

Understanding the NEC3 ECC Contract

A practical handbook

Kelvin Hughes

Routledge
Taylor & Francis Group

LONDON AND NEW YORK

First published 2013
by Routledge
2 Park Square, Milton Park, Abingdon, Oxon OX14 4RN

Simultaneously published in the USA and Canada
by Routledge
711 Third Avenue, New York, NY 10017

Routledge is an imprint of the Taylor & Francis Group, an informa business

This publication presents material of a broad scope and applicability.
Despite stringent efforts by all concerned in the publishing process, some
typographical or editorial errors may occur, and readers are encouraged
to bring these to our attention where they represent errors of substance.
The publisher and author disclaim any liability, in whole or in part,
arising from information contained in this publication. The reader is
urged to consult with an appropriate licensed professional prior to
taking any action or making any interpretation that is within the
realm of a licensed professional practice.

Trademark notice: Product or corporate names may be trademarks or
registered trademarks, and are used only for identification and
explanation without intent to infringe.

British Library Cataloguing in Publication Data
A catalogue record for this book is available from the British Library

Library of Congress Cataloging in Publication Data
Hughes, Kelvin.
    Understanding the NEC3 ECC contract : a practical handbook / Kelvin Hughes.
    p. cm.
    Includes bibliographical references and index.
    1. Civil engineering contracts—Great Britain. 2. Construction contracts—Great Britain.
    3. NEC Contracts. I. Title.
    KD1641.H836 2012
    343.4107'869—dc23
    2011053138

ISBN: 978–0–415–61496–2 (pbk)
ISBN: 978–0–203–82931–8 (ebk)

Typeset in Goudy
by Swales & Willis Ltd, Exeter, Devon

Printed and bound by CPI Group (UK) Ltd, Croydon, CR0 4YY

Contents

3   **Time**   85

4   **Testing and Defects**   116

List of figures

Preface

I have been involved with the NEC family of contracts since 1995, and in that time have advised on numerous projects using the family of contracts, and carried out over 1,200 NEC-based training courses in the UK and abroad.

I was also Secretary of the NEC Users Group from 1996 to 2006, providing support to users and potential users of the contracts including seminars and workshops, during which time I also ran the Users Group helpline answering queries on the NEC contract from members of the Group.

During this involvement with the NEC, particularly with the Engineering and Construction Contract (ECC), I have always felt there was a need for a practical manual on the contract and its use including worked examples and illustrations – essentially a training course within a book, including, for example, what does an activity schedule look like, how do time risk allowances and float work, how do you assess compensation events for omissions of work, etc, which are not covered either by the Guidance Notes, or by other text books.

The Guidance Notes that accompany the contract were written by the contract drafters, whom I have known and always admired, and whilst the Guidance Notes are comprehensive and well written, I have always felt that they do not adequately act as a practical handbook. Various books that are available have in many cases been written by people who may not be actively involved with NEC contracts and training people to use them.

This book is intended to fill the gap where the Guidance Notes leave off, and cover issues that may not have been considered or covered within the official publications.

It is intended to be of benefit to professionals who are actually using the contract, but also to students who need some awareness of the contract as part of their studies.

Whilst a number of practitioners still use the NEC 2nd Edition Engineering and Construction Contract, the book is primarily aimed at giving guidance to NEC3 users, though the structure and content of the contracts are very similar and to that end much of the advice given in this book is of use to all users.

As I have always viewed the NEC3 contracts as a manual of good practice as well as a contract, particularly in terms of the use of its Main and Secondary Options, risk management through the early warning system, clear requirement for a programme, and a disciplined procedure for change management, I have deliberately included within each chapter an overview of the subject area covered by the ECC, and discussed how other contracts deal with the issues, as well as detailing the NEC3 provisions, and hope that will give the readership a more rounded view of good project management principles. Hopefully this will also be useful to student readers.

Readers may note the absence of case law within the text of the book. This is a deliberate policy on my part, for three reasons.

First, I am not a lawyer, my background being in senior commercial positions with major building contractors, so I felt, and readers may concur, that I was not qualified to quote and to attempt a detailed commentary on any case law.

Second, there has actually been very little case law on the NEC contracts since they were first launched.

Third, and probably the most important, as a contracts consultant with significant overseas experience of all contracts including the NEC, it was always my intention that the book should attract an international readership. NEC was always conceived as an international contract, so including UK case law would probably limit it to a UK readership.

I have included within the text many examples, as I have found when running NEC3 training courses that delegates understand principles much better if they can be given illustrated and worked examples, and where numbers are involved using real calculations.

Throughout I have also endeavoured to maintain the NEC drafting policy of using capital letters for defined terms eg 'Plant', 'Materials', 'Completion', 'Defects' and 'Prices' etc. I have also used capital letters for some terms (eg. 'Main and Secondary Options', 'Defects Date', 'Works' etc) which are in some cases italicised within the contract, to give them prominence within the text.

My personal motivation for writing a book is to collate into one volume a significant quantity of material which I have gathered over the years and to share with others a substantial knowledge and experience of the contract.

Kelvin Hughes
December 2011

Acknowledgements

I would like to extend my sincere thanks and gratitude to the following people who have supported me in writing this book:

- Roger Lewendon for agreeing to read my drafts, for his invaluable advice on contractual issues, and for his friendship, wisdom and guidance on life itself!
- Lesley, the love of my life, who gives me the time, the inspiration and the support to fulfil my life's ambitions!
- my parents, Maureen and Dennis, who made me who I am, and I hope in return I have made them proud.

Introduction

0.1 Background to NEC3

The history of the New Engineering Contract stems from 1985, when the Council of the Institution of Civil Engineers (ICE) approved a recommendation from its Legal Affairs Committee 'to lead a fundamental review of alternative contract strategies for civil engineering design and construction with the objective of identifying the needs for good practice'.

There had been a concern for many years that the policy of construction contracts being sector specific, ie building, civil engineering, process, etc and using a single procurement route was outdated. In addition, there had been concerns for many years about issues such as post-Completion prolonged claims negotiations, and the time taken to agree final accounts.

In 1986 the specification for the new contract was submitted to the Legal Affairs Committee, and in 1991 the first consultation document was issued with consultation lasting a period of two years. It is perhaps worth mentioning that all of the successive NEC contracts have gone through a consultation process, the drafters inviting comments by the NEC Users Group, so that potential users can contribute to the proposed new contracts.

The New Engineering Contract (now NEC3), is now being used substantially within the construction industry, and by many Employers in the public and private sectors in several countries. There are now several national and international flagship projects, and tens of thousands of other projects which have been, and are being, built using the contract, with most Employers having reported that the contract gives far greater control of time, cost and quality issues combined with greatly improved relationships between the contracting parties.

In July 1994, Sir Michael Latham's report *Constructing the Team*[1] recommended that the then New Engineering Contract should be adopted by Employers as the standard form of contract to be used in both the private and public sectors rather than using bespoke forms of contract, and suggested that it should become a national standard contract across the whole of engineering and construction work generally. He stated that endlessly refining existing conditions of contract would not solve adversarial problems, a set of basic principles is required on which modern contracts can be based, a complete family of interlocking documents is also required.

He also stated that the New Engineering Contract (NEC) fulfils many of these principles and requirements, but that changes to it would be desirable and the matrix was not yet complete.

Latham also stated that the most effective form of contract should include:

1. A specific duty for all parties to deal fairly with each other and in an atmosphere of mutual co-operation.
2. Firm duties of teamwork, with shared financial motivation to pursue those objectives.
3. A wholly interrelated package of documents suitable for all projects and any procurement route.
4. Easily comprehensible language and with guidance notes attached.
5. Separation of the roles of contract administrator, project or lead manager and adjudicator.
6. A choice of allocation of risks, to be decided as appropriate to each project.
7. Where variations occur, they should be priced in advance, with provision for independent adjudication if agreement cannot be reached.
8. Provision for assessing interim payments by methods other than monthly valuation, ie milestones, activity schedules or payment schedules.
9. Clearly setting out the period within which interim payments should be made, failing which there should be an automatic right to compensation involving payment of interest.
10. Providing for secure trust fund routes of payment.
11. While avoiding conflict, providing for speedy dispute resolution by an impartial adjudicator/referee/expert.
12. Providing for incentives for exceptional performance.
13. Making provision where appropriate for advance mobilisation payments.

The NEC family now provides for all of Latham's recommendations.

In June 2005, NEC3, the current generation of NEC contracts, was launched as a complete review and update of the whole NEC family, including the introduction of two new contracts, the Term Service Contract and the Framework Contract. Since that date, the Supply and Supply Short Contracts have also joined the family.

Alongside the launch of NEC3, the Office of Government Commerce (OGC) announced that after extensive research, they identified NEC3 as the contract which should be adopted by private and public Employers in commissioning their projects.

This was reflected on the inside front cover of the NEC3 contracts when they launched in June 2005, which stated:

> OGC advises public sector procurers that the form of contract used has to be selected according to the objectives of the project, aiming to satisfy the Achieving Excellence in Construction (AEC) principles. This edition of the NEC (NEC3) complies fully with the AEC principles. OGC recommends the use of NEC3 by public sector construction procurers on their construction projects.[2]

This note was changed in the 2010 reprint to:

> The Construction Clients' Board (formerly Public Sector Construction Clients Forum) recommends that public sector organisations use the NEC3 contracts when procuring construction. Standardising use of this comprehensive suite of contracts should help to deliver efficiencies across the public sector and promote behaviours in line with the principles of Achieving Excellence in Construction.[3]

It should be stressed that NEC3 is the only contract to have received this recommendation to date.

Whilst articles written about NEC tend to give prominence to many of the big 'flagship' projects carried out using the contracts, of much more significance are the very many more modest size and value projects being procured. It is also in this area that many of the problems have arisen in implementation of the contract, where perhaps less experienced practitioners may be involved, some of which have hopefully been addressed within this book.

The original objectives of the NEC contracts, and more specifically the Engineering and Construction Contract (ECC), which is the specific topic of this book, were to make improvements under three headings:

(i) Flexibility

The contract should be able to be used:

- for any engineering and/or construction work containing any or all of the traditional disciplines such as civil engineering, building, electrical and mechanical work, and also process engineering.
  Previously, contracts had been written for use by specific sectors of the industry, eg ICE civil engineering contracts, JCT building contracts, IChemE process contracts;
- whether the Contractor has full, some or no design responsibility.
  Previously, most contracts had provided for portions of the work to be designed by the Contractor, but with separate design and build versions if the Contractor was to design all or most of the Works;
- to provide all the normal current options for types of contract such as lump sum, remeasurement, cost reimbursable, target and management contracts.
  Previously, contracts were written, primarily as either lump sum or remeasurement contracts, so there was no choice of procurement method when using a specific standard form;
- to allocate risks to suit each particular project.
  Previously, contracts were written with risks allocated by the contract drafters, and one had to be expert in contract drafting to amend the conditions to suit each specific project on which it was used;
- anywhere in the world.

Previously, contracts were written with country-specific clauses, which made it difficult to use them in other countries. Any country-specific NEC clauses are included as Secondary Options eg Option Y(UK)2 and Y(UK)3.

To date, the NEC contracts have been used for projects as widely diverse as airports, sports stadiums, water treatment works, housing projects, in many parts of the world and even research projects in the Arctic!

(ii) Clarity and simplicity

- The contract is written in ordinary language and in the present tense.
- As far as possible, NEC only uses words which are in common use so that it is easily understood, particularly where the user's first language is not English. Previously obscure words such as 'whereinbeforesaid', 'hereinafter' and 'aforementioned' were commonplace in contracts! It has few sentences that contain more than 40 words and uses bullet points to subdivide longer clauses.
- The number of clauses and the amount of text are also less than in most other standard forms of contract and there is an avoidance of cross-referencing found in more traditional standard forms.
- It is also arranged in a format which allows the user to gain familiarity with its contents, and required actions are defined precisely thereby reducing the likelihood of disputes.
- Finally, subjective words like 'fair' and 'reasonable' have been used as little as possible as they can lead to ambiguity, so more objective words and statements are used.

Some critics of NEC have commented that the 'simple language' is actually a disadvantage as certain clauses may lack definition and there are certain recognised words that are commonly used in contracts. There is very little case law in existence with NEC contracts, and whilst adjudications are confidential and unreported, anecdotal evidence suggests that there does not appear to be any more adjudications with NEC contracts than any other, which would tend to suggest that the criticism may be unfounded.

(iii) Stimulus to good management

This is perhaps the most important objective of the ECC in that every procedure has been designed so that its implementation should contribute to, rather than detract from, the effective management of the work. In order to be effective in this respect, contracts should motivate the parties to proactively want to manage the outcome of the contract, not just to react to situations. NEC intends and requires the parties to be proactive and not reactive. It also requires the parties and those that represent them to have the necessary experience (sadly often lacking) and to be properly trained so that they understand how the NEC works.

The philosophy is founded on two principles:

1. 'Foresight applied collaboratively mitigates problems and shrinks risk.'
2. 'Clear division of function and responsibility helps accountability and motivates people to play their part.'

Examples of foresight within the ECC are the early warning and compensation event procedures, the early warning provision requiring the Project Manager and the Contractor each to notify the other upon becoming aware of any matter which could have an impact on price, time or quality.

A view held by many Project Managers is that an early warning is something that the Contractor would give, and is an early notice of a 'claim'. This is an erroneous view as, first, early warnings should be given by either the Project Manager or the Contractor, whoever becomes aware of it first, the process being designed to allow the Project Manager and Contractor to share knowledge of a potential issue before it becomes a problem, and, second, early warnings should be notified regardless of whose fault the problem is, it is about raising and resolving the problem, not compensating the affected party.

The compensation event procedure requires the Contractor to submit, within three weeks, a quotation showing the time and cost effect of the event. The Project Manager then responds to the quotation within two weeks, enabling the matter to be properly resolved close to the time of the event rather than many months or even years later.

The programme is also an important management document with the contract clearly prescribing what the Contractor must include within his programme and requiring the Project Manager to 'buy into it' by formally accepting (or not accepting) the programme. The programme is then defined as the 'Accepted Programme'.

In total the ECC is designed to provide a modern method for Employers, Contractors, Project Managers, Designers and others to work collaboratively and to achieve their objectives more consistently than has been possible using other traditional forms of contract. People will be motivated to play their part in collaborative management if it is in their commercial and professional interest to do so.

Uncertainty about what is to be done and the inherent risks can often lead to disputes and confrontation but the ECC clearly allocates risks and the collaborative approach will reduce those risks for all the parties so that uncertainty will not arise.

0.2 Essential differences between the ECC and other forms of contract

The ECC is part of the NEC3 matrix of contracts – Engineering and Construction Contract and Subcontract, Engineering and Construction Short Contract and Short Subcontract, Professional Services Contract, Term Service and Term Service Short Contract, Supply and Supply Short Contract, Framework Contract, and the Adjudicator's Contract – allowing all parties, whatever the project or service to be provided, to work under similar conditions.

- *Flexibility of use*: The ECC is not sector specific, in terms of it being a building, civil engineering, mechanical engineering, process contract, etc as with other contracts, it can be used for any form of engineering or construction. This is particularly useful where a major project such as an airport or a sports stadium can be a combination of building, civil engineering and major mechanical and electrical elements.
- *Flexibility of procurement*: The ECC's Main and Secondary Options, together with flexibility in terms of Contractor design, allow it to be used for any procurement method whether the Contractor is to design all, none or part of the Works. Again, other contracts do not offer this flexibility.
- *Early warning*: The ECC contains express provisions requiring the Contractor and the Project Manager to notify and, if required, call a 'risk reduction' meeting when either becomes aware of any matter which could affect price, time or quality. Few contracts have this express requirement.
- *Programme*: There is a clear and objective requirement for a detailed programme with method statements and regular updates which provides an essential tool for the parties to manage the project and to notify and manage the effect of any changes, problems, delays, etc. Whilst other contracts contain programme requirements, they do not deal with them in the same detail, and one could argue that they probably do not properly recognise the importance of a programme.
- *Compensation events*: This procedure is unique to NEC and requires the Contractor to price the time and 'Defined Cost' effect of a change within three weeks and for the Project Manager to respond within two weeks. There is therefore a 'rolling' Final Account with early settlement and no later 'end of job' claims for delay and/or disruption. It is also more beneficial for the Contractor in terms of his cash flow as the Contractor is paid agreed sums rather than reduced 'on account' payments, which are subject to later agreement and payment.
- *Disputes*: The contract encourages better relationships and there is far less tendency for disputes because of its provisions. If a dispute should arise there are clear procedures as to how to deal with it, ie adjudication, tribunal.

0.3 Arrangement of the ECC

The ECC includes the following sections:

(i) Core clauses

1. General
2. The Contractor's main responsibilities
3. Time
4. Testing and Defects
5. Payment
6. Compensation events
7. Title

8. Risks and insurance
9. Termination.

(ii) *Main Option clauses*

- Option A: Priced contract with activity schedule
- Option B: Priced contract with bill of quantities
- Option C: Target contract with activity schedule
- Option D: Target contract with bill of quantities
- Option E: Cost reimbursable contract
- Option F: Management contract.

(iii) *Dispute resolution*

- Option W1: Dispute resolution procedure (used unless the Housing Grants, Construction and Regeneration Act 1996* applies).
- Option W2: Dispute resolution procedure (used when the Housing Grants, Construction and Regeneration Act 1996* applies).

(iv) *Secondary Option clauses*

- Option X1: Price adjustment for inflation
- Option X2: Changes in the law
- Option X3: Multiple currencies
- Option X4: Parent company guarantee
- Option X5: Sectional Completion
- Option X6: Bonus for early Completion
- Option X7: Delay damages
- Option X12: Partnering
- Option X13: Performance bond
- Option X14: Advanced payment to the Contractor
- Option X15: Limitation of the Contractor's liability for his design to reasonable skill and care
- Option X16: Retention
- Option X17: Low performance damage
- Option X18: Limitation of liability
- Option X20: Key Performance Indicators
- Option Y(UK)2: The Housing Grants, Construction and Regeneration Act 1996
- Option Y(UK)3: The Contracts (Rights of Third Parties) Act 1999
- Option Z: Additional conditions of contract.

NB Options X8 to X11, X19 and Y(UK)1 are not used.**

---

* See amendment regarding the Local Democracy, Economic Development and Construction Act 2009 described later in this chapter.

** These Options appear in other NEC3 contracts.

(v) Schedules of Cost Components

(vi) Contract Data

Other documents to be used with the ECC include:

- the Works Information;
- the Site Information;
- the Accepted Programme;
- other documents resulting from choosing various Secondary Options, eg Performance Bond.

Depending on the choice of Main Options the documents may also include:

- an activity schedule (Option A or C); or
- a bill of quantities (Option B or D).

0.4 The Main and Secondary Options

The Main Options

The six Main Options A to F enable Employers to select a procurement strategy and payment mechanism most appropriate to the project and the various risks involved. Chapter 5 includes more details on payment mechanisms and the contractual requirements under each Main Option. Essentially, the Main Options differ in the way the Contractor is paid.

Whilst many traditional contracts are based on bills of quantities, there has been a movement away from the use of traditional bills and towards payment arrangements such as milestone payments and activity schedules, with payment based on progress achieved, rather than quantity of work done. There is also an increasing use of target cost contracts which has been encouraged by the increasing use of partnering arrangements, the better sharing of risk and also the continued growth of NEC contracts which provide for target options. To that end, it is perhaps not surprising that in a survey carried out by the RICS[4] Options A and C were found to be the most regularly used NEC3 Main Options.

Whilst there was only a small representative sample of 106 contracts, the findings in terms of how often each Option was selected were:

- Option A = 33 per cent
- Option B = 7 per cent
- Option C = 55 per cent
- Option D = 0 per cent
- Option E = 2 per cent
- Option F = 1 per cent

In 2 per cent of the contracts surveyed, the Engineering and Construction Short Contract was selected.

Once the procurement strategy has been decided, the Main and Secondary Options can be selected to suit that strategy.

The Main Options are:

- Option A: Priced contract with activity schedule
- Option B: Priced contract with bill of quantities
- Option C: Target contract with activity schedule
- Option D: Target contract with bill of quantities
- Option E: Cost reimbursable contract
- Option F: Management contract.

- Options A and B are priced contracts in which the risks of being able to carry out the work at the agreed Prices are largely borne by the Contractor.
- Options C and D are target contracts in which the Employer and Contractor share the financial risks in an agreed proportion.
- Options E and F are two types of cost reimbursable contract in which the financial risks of being able to carry out the work are largely borne by the Employer.

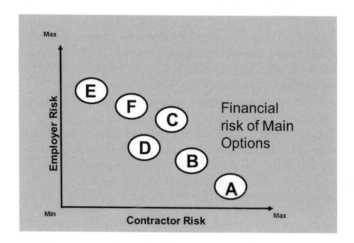

Figure 0.1 Financial risk of Main Options A–F

Option A: Priced contract with activity schedule

Procurement using specification and drawings or activity schedules is still the way the largest number of projects, particularly those lower in value, are procured. In addition, many projects are procured on a design and build basis. In that respect, Option A has always been the most commonly selected Main Option.

Option A is normally used where the Employer knows exactly what he wants, and is able to clearly define it through the Works Information, which would mainly comprise drawings and specifications, but can alternatively be a performance specification, where the Contractor is to design the Works to meet specific performance objectives.

Whilst the ECC is appropriate for all forms of engineering and construction, Option A is particularly appropriate for building projects where the work can more often be clearly defined.

The activity schedule is a list of activities, normally prepared by the Contractor, and can be used where the Employer has provided all the Works Information, or where the Contractor is to design all or part of the Works based on some form of performance criteria, and therefore there is also Works Information provided by the Contractor.

When the Contractor has priced the activity schedule, the lump sum for each activity is the Price to be paid by the Employer in the assessment following Completion of that activity.

Option A is therefore a stage payment contract and as payment is linked to Completion of activities, the Contractor must plan and carry out his work effectively with the cash flow requirements for both parties being clearly visible. Administration of payment aspects is therefore fairly simple and transparent.

Establishing and taking off the quantities of work involved to achieve the Completion of each activity is the responsibility of the tendering Contractors. The Price for each activity is in effect a lump sum for carrying out and completing that activity and must include everything necessary to complete the activity. The sum of the tendered lump sums for each of the activities is the tendered Price (the Total of the Prices) for the whole of the Works.

Option A is ideally suited to Contractor's design but can be used for Employer's design or split design responsibility.

Under Option A, the Contractor prepares and submits an activity schedule with his tender. The Total of the Prices in the activity schedule is the Contractor's Price for Providing the Works. The contract requires that the activities on the activity schedule relate to operations on each programme.

It is important to recognise that the activity schedule has two primary functions:

1. It shows how the tendering Contractors have built up their Price and can therefore be used as part of the tender assessment process.
2. It is used to calculate the Price for Work Done to Date.

The Employer does not provide a bill of quantities with Option A, the taking off of the quantities of work involved to achieve the Completion of each activity and the pricing of the activity being the responsibility of the tendering Contractors.

Typical Activity Schedule		£
Small Industrial Unit		
Preliminaries		
Set up Site Compound		12564
Time-based	Month 1	7900
Preliminaries	Month 2	7900
	Month 3	7900
	Month 4	7900
	Month 5	7900
	Month 6	7900
	Month 7	7900
	Month 8	7900
	Month 9	4230
Clear Site Compound		7231
Bulk Excavation	GL 1 – 10	46358
	GL 10 – 20	61327
Main Building		
Reduce Level Excavation		23458
Ground Beams	Formwork	12453
	Reinf	15555
	Conc	12342
Grd Floor Const	Formwork	18239
	Reinf	17298
	Conc	31067
Steel Frame	Delivery to Site	73451
	Erection – GL 1 – 6	12233
	Erection – GL 6 – 12	12233
	Erection – GL 12 – 18	12233
Roof Covering	GL 1 – 9	24986
	GL 9 – 18	28932
External Cladding	GL A1 – A9	22356
	GL A9 – A18	24376
	GL A18 – G18	28765
	GL G18 – G9	24376
	GL G9 – G1	22356
	GL G1 – A1	28765
Windows		14329
Internal Walls		23325
Joinery (1st Fix)		11987
M&E (1st Fix)	Mechanical	43259
	Electrical	28561
Wall Finishings		33456
Joinery (2nd Fix)		28979
Ceiling Finishes		19267
M&E (2nd Fix)		48765
Floor Finishings		7863
External Works		
SW Drainage		23876
Kerbs & Pavings		21985
Tarmacadam		12879
White Lining		11435
	Total of the Prices:	970350

Figure 0.2 Typical activity schedule: Options A and C

There are certain advantages to the Employer using Option A:

- There is no requirement for a bill of quantities to be prepared for tendering Contractors to price. This will save the Employer time and money at the pre-tender stage.
- The Contractor holds the risk of inaccuracies in the quantities, or missing an item of work stated within the Works Information. However, it must be reinforced that the Employer holds the risk of inaccuracies and omissions within the Works Information itself.
- The assessment of the Price for Work Done to Date is easier and quicker than with the other Main Options. There is no requirement to remeasure the work and apply quantities to rates and Prices to calculate the amount due to the Contractor.
- Programme and activity schedule preparation are linked as integrated activities which would tend to lead to a more comprehensive tender.
- Payment is linked to Completion of an activity or group of activities related back to the programme, so cash flow requirements for both parties are more visible.
- In order to receive payment, the Contractor has to complete an activity by the assessment date, so he has to price and programme realistically and accurately, and is motivated to keep to that programme during construction.
- The assessment of the effects of compensation events is related back to the activity schedules. Any change in resources or methods associated with an activity can be compared with those stated in the Accepted Programme before the compensation event occurred.

The disadvantages to the Employer are:

- As there is no specific document that all the tendering Contractors price, it is difficult to assess tenders on a 'like-for-like' basis. Some Employers cite that as an advantage, saying that they do not require the facility to check tenders on a line-by-line basis anyway!

  It is not uncommon, to assist the assessment of tenders, for Employers to issue templates for the Contractor to price covering the main construction elements as part of the tender with the tendering Contractors then inserting the activities under within each element. In that way, each element will have been priced on a 'like-for-like' basis, but the activities within each element may have been described and priced differently by each tenderer.

  If a template is used, Employers should be wary of them being too prescriptive and almost being seen by tenderers as a bill of quantities.
- The tendering Contractors have to spend time, and money, in taking off the quantities of work in order to establish and populate the activity schedule. In order to reduce the tenderers' efforts in producing their own

quantities, some Employers issue quantities to tenders, but these must be stressed as for indicative purposes only, and tenderers should not rely on those quantities.

• The Employer holds the risk of clearly defining the Works Information in order that the Contractor can prepare the activity schedule.

When activity schedules are used, payment is linked to Completion of activities, the Contractor must plan and carry out his work effectively with the cash flow requirements for both parties being clearly visible. Administration of payments is therefore fairly simple.

This Option is ideally suited to Contractor's design but can be used for Employer's design or divided design responsibility.

Option B: Priced contract with bill of quantities

Option B is essentially a remeasurement contract based on a bill of quantities prepared by the Employer with the Contractor pricing the items in the bill of quantities including matters which are at his risk. The Contractor is then paid for the quantity completed to date at the rates and Prices in the bill of quantities.

In the early days of the New Engineering Contract, Employers would tend to select Option B initially as it mirrored the traditional bills of quantities-based contracts they were using at the time. Once Employers became more experienced and confident in the use of the contract, they tended to move towards other Main Options.

It is important that, if selecting a remeasurement contract, one should consider the merits of preparing and using bills of quantities against other procurement methods. An Employer should not select remeasurement just because in the past, with other forms of contract, he has always begun the procurement process by producing the drawings and specification then measuring the work! Again, this comes back to a procurement strategy consideration.

It must be noted that changes which are compensation events are inserted into the bill of quantities as lump sums rather than on a remeasurement basis (see Chapter 6).

This Option is normally used where the Employer knows what he wants and is able to clearly define it through the Works Information, and measure it within the bills of quantities, but there are likely to be changes in the quantities, that may or may not be considered as compensation events.

Although there has been a general decline in use of bills of quantities in the past 25 years, the Option is still quite widely used on civil engineering projects where the final quantity of excavations, filling materials, etc cannot be reasonably forecast with great accuracy.

Option B does not work well when there is a significant amount of Contractor's design, in this case, a lump sum contract such as Option A or perhaps Option C should be used rather than remeasurement, as the Contractor can then carry out initial design calculations, price the work and include the various design stages within his activity schedule rather than the Employer measuring the work, which would not be practical.

A bill of quantities comprises a list of work items and quantities prepared by the Employer and priced by the Contractor. Standard methods of measurement are published (eg SMM7 and CESMM3 or local equivalents) which state the items to be included and how the quantities are calculated.

Confusion can sometimes arise as to whether the tendering Contractors should price the bills of quantities or the Works Information. The Contractor Provides the Works in accordance with the Works Information (Clause 20.1), therefore it is assumed that he programmes the Works in accordance with the Works Information. Clause 31.2 (Programme) refers to 'Providing the Works' and 'Works Information'.

Information in the bill of quantities is not Works Information or Site Information (Clause 55.1). It does not tell the Contractor what he has to do, but it must clearly be a fair representation of the Works to allow the Contractor to accurately price the Works. Although a Contractor will often use the quantities within the bill of quantities as a guide to calculate time scales, the bill is used as a basis for inviting and assessing tenders, so it is a pricing and payment document, ie it deals with money. It is reasonable for the Contractor to presume that the quantities are a fair representation of what is included in the Works Information.

If there is an inconsistency between the documents which form part of the contract either the Project Manager or the Contractor notifies the other as soon as either becomes aware that it exists (Clause 17.1) and the Project Manager corrects it. This may then be addressed as a compensation event under Clause 60.6, with Clause 60.7 stating that in assessing a compensation event resulting from the correction of an inconsistency between the bill of quantities and another document, the Contractor is assumed to have taken the bill of quantities as correct. This is in respect of the Contractor's Price, Completion Date and programme, so the Price and, if appropriate, the Completion Date may then be changed to accommodate the correction.

Clearly, if the Contractor notices the error during the tender period then he should raise it as a query, then the matter can be dealt with and all tenderers informed, however, while the Contractor should notify if and when he finds an error, he is not under any obligation to look for the error!

Option C: Target contract with activity schedule

Option C is a target contract based on an activity schedule, again as with Option A normally prepared by the Contractor. It must be emphasised that the activity schedule is not used for payments in the same way as with an Option A contract. In Option C, the activity schedule acts as a guide when assessing tenders, but payments are based on Defined Cost incurred, not Completion of activities.

Target contracts are a development of cost reimbursable contracts and are a combination of a lump sum price and open book cost reimbursable payments.

The advantage of target contracts is that the parties initially have certainty of Price, with the Contractor being incentivised to make cost savings for the benefit of the Employer and himself.

This Option is normally used where the Employer knows what he wants and is able to clearly define it through the Works Information, so that the Contractor can price and prepare the activity schedule, but he sees a benefit in sharing risk and opportunity with the Contractor thereby encouraging collaboration. Financial risks are shared proportionally between the Employer and the Contractor through the Contractor's share percentages. Many wrongly call Option C the 'Partnering Option' although it does lend itself to the principles of partnering, collaborative working and associated risk/opportunity sharing. Coincidentally, there is a Partnering Option X12 which may have given rise to the confusion. This Option is described in detail later in this chapter.

The Contractor tenders a Price and includes an activity schedule in the same way as he would under an Option A contract. This Price, when accepted, is then referred to as the 'target'. The Contractor also tenders his percentage(s) for Fee. The original target is referred to as the 'Total of the Prices at the Contract Date'.

- The target Price includes the Contractor's estimate of Defined Cost plus other costs, overheads and profit to be covered by his Fee.
- The Contractor tenders his Fee in terms of percentages to be applied to Defined Cost.
- During the course of the contract, the Contractor is paid Defined Cost plus the Fee.
- The target is adjusted for compensation events and also for inflation (if Option X1 is used).
- On Completion, the Project Manager assesses the Contractor's share in accordance with Clause 53.1 which, it has to be said, is at best a confusing clause, though the Guidance Notes clarify the clause! The Contractor then pays or is paid his share of the difference between the final total of the Prices and the final Payment for Work Done to Date according to a formula stated in the Contract Data. This motivates the Contractor to decrease costs. Many refer to this sharing of risk and opportunity as 'pain and gain'. It often comes as a surprise to NEC users that the terms 'pain and gain' do not appear anywhere within the NEC contracts, nor do the terms 'target price' or 'target cost'!

Under Options C, D, E and F the Contractor is required to submit forecasts at the intervals stated in the Contract Data of the Defined Cost for the Works, which advises the Employer of his potential outturn cost.

Option D: Target contract with bill of quantities

Option D is a target contract based on a bill of quantities. Again, as with Option B, the bills of quantities are prepared by the Employer.

This Option is normally used where the Employer knows what he wants and is able to define it clearly through the Works Information, and measure it within the bills of quantities, but there are likely to be changes in the quantities, that may or may not be considered as compensation events.

The Contractor tenders a Price based on the bills of quantities. This Price, when accepted, is then referred to as the 'target'. The original target is referred to as the Total of the Prices at the Contract Date.

The assessment of payments is again the same as for an Option C contract:

- The target Price includes the Contractor's estimate of Defined Cost plus other costs, overheads and profit to be covered by his Fee.
- The Contractor tenders his Fee in terms of percentages to be applied to Defined Cost.
- During the course of the contract, the Contractor is paid Defined Cost plus the Fee.
- The target is adjusted for compensation events and also for inflation (if Option X1 is used).
- On Completion, the Contractor is paid (or pays) his share of the difference between the final Total of the Prices and the final Payment for Work Done to Date according to a formula stated in the Contract Data. If the final Payment for Work Done to Date is greater than the final Total of the Prices, the Contractor pays his share of the difference.
- As with Option B the target is generated as a remeasurement, based on the bill of quantities, though changes which are compensation events are inserted into the bill of quantities as lump sums rather than on a remeasurement basis (see Chapter 6).

Option E: Cost reimbursable contract

Option E is a cost reimbursable contract with the Contractor being reimbursed Defined Cost plus the Fee.

It should be used:

- where the scope of the Works is uncertain, eg some refurbishment projects;
- where extreme flexibility is required, eg for enabling work;
- where a high level of Employer involvement is envisaged;
- for emergency work;
- where trials or work of an experimental nature are carried out.

The Option allows development of the design as the Works proceed and permits maximum flexibility in allocation of design responsibility.

A cost reimbursable contract should be used where the definition of the work to be done is inadequate for a lump sum Price, or even as a basis for a target Price, and yet an early start is required.

A criticism of a cost reimbursable contract such as Option E is that it gives the Contractor very little if any incentive to reduce costs; however, if the Employer cannot clearly define the work to be done so that the Contractor can price it, but he wishes the work to be carried out, the Contractor cannot be expected to take financial risks. The Contractor carries minimum risk and is reimbursed his Defined Cost plus Fee, subject to any deduction for Disallowed Costs.

Under Option E of the ECC, the Contractor is reimbursed his Defined Cost plus a Fee, basically covering his off-site overheads and profit. This Fee is calculated by applying the Fee percentage, given at tender by the Contractor in Contract Data Part 2, to appropriate Defined Cost. Although Option E is a cost-reimbursable contract, there is an obligation on the Contractor to provide a regular forecast of the Defined Cost for the Works which advises the Employer of his potential outturn cost.

Another criticism of Option E is that the tendering Contractors do not actually price the Works, they just price their Fee percentage together with any appropriate Equipment rates, Working Areas overheads percentage and manufacture and fabrication rates and overheads percentages. In addition, tenderers may be required to notionally agree to a cost plan. This makes it difficult to assess cost-reimbursable tenders, but again the basis of cost reimbursable contracts is that the Employer bears most of the risk.

Option F: Management contract

Management contracting emerged as a procurement method for large projects in the early 1980s, but its use declined in the 1990s and it is now rarely used, but the ECC provides for it through Option F.

The principle with a management contract is that the Contractor focuses on, and is paid for, managing the contract rather than physically building the project and does not normally carry out any construction work himself.

The Contractor's responsibilities under a management contract normally cover:

- provision of Site accommodation, and other common-use facilities;
- managing time-based deliverables such as programme, milestones and project Completion;
- management of cost, normally in conjunction with the Employer's cost consultant;
- management of quality and Defects-related issues;

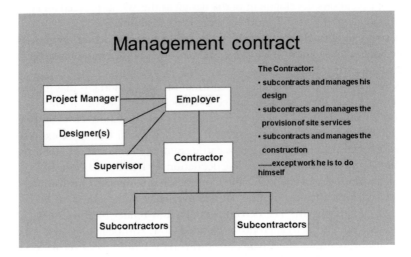

Figure 0.3 Option F Management contract: Contractor's responsibilities

- managing the design, though in most cases he is not responsible for carrying out the design, 'design and manage' is a procurement method where the Contractor normally has full design responsibility;
- assembling, tendering and managing Subcontractor packages.

The amounts paid to the Subcontractors are reimbursed to the Contractor, plus a Fee percentage.

The Contractor holds very little risk under this procurement method; he is reimbursed the Subcontractor costs and any other costs (less any Disallowed Costs) plus his Fee.

The Contractor's work mainly applies to the construction phase though he can be appointed for pre-construction services. All subcontracts are directly with the Contractor.

The ECC specifically states, under Option F, that the Contractor's responsibility is to manage the design that has been assigned to him, the provision of Site services and the construction and installation of the Works, all of which are subcontracted. The Contractor is also required to advise the Project Manager on the practical implications of the design of the Works and on subcontracting arrangements. He is also required to prepare forecasts of the total Defined Cost for the whole of the Works in consultation with the Project Manager at intervals stated in the Contract Data.

The Contractor tenders his Fee percentage, and in some cases may be required to agree that an estimated Total of the Prices produced by the Employer or his representative is reasonable, though it must be stressed that the Contractor is reimbursed costs, and is therefore not liable for carrying out the work for those Prices.

Management contracts are generally suitable:

- where there is a need to coordinate a number of Subcontractors and Suppliers;
- when the Employer does not have sufficient capability to manage the project himself;
- when the time scale of the project is tight requiring an early start of construction;
- where the scope of the project may not be fully developed. As the scope is developed and construction progresses, successive subcontracts can be awarded, which need to be managed, together with their interfaces.

Again, as with Option E, the tendering Contractors do not actually price the Works, they just price their Fee percentage, with further rates or percentages as with the other Options as the Schedule of Cost Components do not apply. Again, this makes it difficult to assess management contract tenders, but again the basis of this type of contract is that the Employer bears most of the risk.

Secondary Options

Option X1: Price adjustment for inflation (used only with Options A, B, C and D)

The Employer should make the decision at the time of preparing the tender documents as to whether inflation for the duration of the contract is to be:

- the Contractor's risk – in which case he should not select Option X1; or
- the Employer's risk – in which case, he should select Option X1.

The default within the ECC is that the contract is 'fixed price' in terms of inflation, ie the Contractor has priced the work to include any inflation he may encounter during the period of carrying out the contract. If Option X1 is chosen, the Prices are adjusted for inflation as the work progresses, by means of a formula.

Note that under Options C and D, whether Option X1 is selected or not, the Price for Work Done to Date is based on the current cost at the time it is incurred. In the case of Options C and D, Option X1 then adds the price adjustment to the Total of the Prices (the target).

Option X1 is not applicable to Options E and F as the Employer again pays Defined Cost at the time that it is incurred.

The key components of the formula are:

- the 'Base Date Index' (B) is the latest available index before the base date;

- the 'Latest Index' (L) is the latest available index before the assessment date of an amount due;
- the 'Price Adjustment Factor' is the total of the products of each of the proportions stated in the Contract Data multiplied by $(L - B)/B$ for the index linked to it.

Under Options A and B, the amount due includes an amount for price adjustment which is the sum of:

- the change in the Price for Work Done to Date since the last assessment of the amount due multiplied by the Price Adjustment Factor for the date of the current assessment;
- the amount for price adjustment included in the previous amount due; and
- correcting amounts, not included elsewhere, which arise from changes to indices used for assessing previous amounts for price adjustment.

Example (for simplicity ignoring correcting amounts):

Option A

The change in the Price for Work Done to Date = £50,000

The Base Date Index (B) = 280.0

The Latest Index (L) = 295.5

The Price Adjustment Factor is therefore $(L - B)/B$

= (295.5–280.0)/280.0

= 0.055

Inflation since the base date is therefore 5.5%

The amount due under Clause X1 is therefore £50,000 × 0.055 = £2,750

Under Options C and D, the amount due includes an amount for price adjustment which is the sum of:

- the change in the Price for Work Done to Date since the last assessment of the amount due multiplied by $(PAF/(1+PAF))$ where PAF is the Price Adjustment Factor for the date of the current assessment; and
- correcting amounts, not included elsewhere, which arise from changes to indices used for assessing previous amounts for price adjustment.

This amount is then added to the Total of the Prices, ie the target.

Note that if Option X1 is chosen, then Defined Cost for compensation events is assessed by adjusting current Defined Cost back to the base date.

Option X2: Changes in the law

As with Option X1, the default is that the contract is 'fixed price' in terms of changes in the law, ie the Contractor has priced the work to include any changes in the law he may encounter during the period of the contract.

If Option X2 is chosen, and a change in the law occurs after the Contract Date the Project Manager notifies the Contractor of a compensation event. The Prices may be increased or reduced in addition to providing for any delay to Completion.

Note that Option X2 refers to a change in the law of the country in which the Site is located, so, for example, a change in the law in another country where goods are being fabricated for delivery to the Site will not be a compensation event.

Option X3: Multiple currencies (used only with Options A and B)

The currency of the contract is stated in Contract Data Part 1. However, Option X3 provides for items or activities to be paid in an alternative currency, the items or activities, the currency and the total maximum payment in this currency to be listed in the Contract Data, beyond which payments are made in the currency of the contract.

The exchange rates, their source and date of publication are also referred to in the Contract Data.

Option X4: Parent company guarantee

This form of guarantee is given by a parent company (or holding company) to guarantee the proper performance of a contract by one of its subsidiaries (the Contractor), who whilst having limited financial resources himself, may be owned by a larger and more financially sound parent company. In most cases, it is the ultimate parent company that provides the guarantee, but sometimes, particularly when the ultimate parent company is in another country, the parent company may just be a company further up the chain within the group, perhaps the national parent who has sufficient assets to provide the required guarantee.

If a parent company guarantee is required, it may either be provided by the Contract Date or within four weeks of the Contract Date.

The parent company guarantee should have an expiry date, which could be Completion of the project, the Defects Date or may even include the 6-, 10- or 12-year limitation period following Completion of the Works to cover any liability for potential latent Defects, the expiry date being defined within the Works Information.

Parent company guarantees are normally used as an alternative to a performance bond (Option X13).

Such a guarantee is cheaper than a performance bond, as the Contractor will normally just charge an administration fee rather than in the case of a performance bond, where the Contractor is actually paying a premium for an insurance policy, but it may give less certainty of redress because it is not supplied by an independent third party so it is dependent on the survival, and the ability to pay, of the parent company.

However, whilst accepting less independence, parent company guarantees for the proper performance of the contract can be more advantageous than bonds. Rather than receiving a fixed amount in compensation, the parent company is normally obliged to either complete the Works in accordance with the Contractor's original obligations on behalf of its subsidiary company, or fund the Completion of the contract by others, so effectively Completion is guaranteed.

In addition, further recompense can be sought for time delays in Completion through the normal clauses incorporated in the contract. In the case of performance bonds the 'guarantee' is that a sum of money is available to at least partly compensate the Employer in the event of the Contractor's default.

Because the financial strength of the parent company may be linked to that of the Contractor, a parent company guarantee will be acceptable only if the parent company (or holding company) is financially strong and its financial resources are largely independent of those of the Contractor. Obviously, if the insolvency is not limited to the Contractor but also relates to the parent group a parent company guarantee will be virtually useless, save for any entitlement under insolvency law.

Option X5: Sectional Completion

If the Employer requires sections of the Works to be completed by the Contractor before the whole of the Works are completed, then Option X5 should be chosen. References in the contract to the Works, Completion and Completion Date will then apply to either the whole of the Works or a section. Contract Data Part 1 should give a description of each section and the date by which it is to be completed.

Option X5 may be selected together with Option X6 (bonus for early Completion) and/or Option X7 (delay damages).

Option X6: Bonus for early Completion

Often early Completion would be of benefit to the Employer, for example the Contractor is building a stadium and early Completion would allow a sporting event or concert to take place, bringing revenue to the Employer.

Option X6 provides for the Employer to pay a bonus to the Contractor for early Completion. The amount of the bonus is stated by the Employer on a 'per day' basis in Contract Data Part 1 and is calculated from the earlier of Completion and when the Employer takes over the Works, until the Completion Date.

Option X7: Delay damages

Delay damages in the ECC are normally referred to in other contracts as liquidated damages.

Delay damages are pre-defined amounts inserted into Contract Data Part 1 and paid or withheld from the Contractor in the event that he fails to complete the Works by the Completion Date. The amount included within the contract for delay damages should be a 'genuine pre-estimate of likely losses', not a penalty.

Many contracts require the contract administrator (Engineer, Architect, etc) to issue some form of certificate confirming that the Contractor failed to complete on time, and also for the Employer to notify the Contractor in writing that he will be withholding the relevant damages, but the ECC Contract provides for the Contractor to pay delay damages at the rate stated in the Contract Data until Completion or the date on which the Employer has taken over the Works, whichever is earlier, without having to certify that the Contractor has defaulted.

Also, many contracts require the contract administrator to value the Works and the Employer then deducts the liquidated damages from the amount due to the Contractor, but the ECC Contract requires the Project Manager to deduct amounts to be paid or retained from the Contractor within his assessment and certificate.

Under Clause X7.3, if the Employer takes over the Works before Completion, the Project Manager assesses the benefit to the Employer of taking over that part of the Works as a proportion of taking over the whole of the Works and the delay damages are reduced in this proportion. Whilst it is probably the most correct way to assess remaining delay damages, the Project Manager having to assess the benefit to the Employer can at best be a subjective exercise and may possibly lead to disputes with the Contractor. Most other contracts state that the amount of liquidated damages is reduced by the same proportion the part that has been taken over bears to the value of the whole of the works, so if a third of the value of the Works has been taken over, the amount of the liquidated damages is reduced by a third.

Options X8 to X11

These Options are not included in the ECC, but as the NEC3 has a common numbering system for Main and Secondary Options, they are found in other members of the NEC3 family. These Options are all found in the NEC3 Professional Services Contract, Option X8 being Collateral Warranty Agreements, Option X9 Transfer of Rights, Option X10 Employer's Agent and Option X11 Termination by Employer.

This common numbering for the Secondary Options provides a consistency sadly lacking in many other families of standard contracts. So, in NEC3, if one of the other contracts has an option for 'price adjustment for inflation' that option will be X1 and the wording will be almost identical. This feature makes the NEC3 contracts very user-friendly, particularly for those who use a number of members of the family and more than one procurement strategy.

Option X12: Partnering

This Option enables a multi-party partnering agreement to be implemented. In this case Option X12 is used as a Secondary Option common to the contract

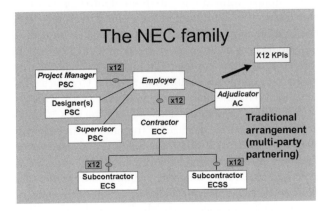

Figure 0.4a Parties' relationships: traditional arrangement

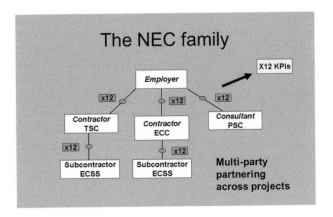

Figure 0.4b Parties' relationships: design and build

which each party has with the body which is paying for the work. The parties together make up the partnering team.

The content of Option X12 is derived from the 'Guide to Project Team Partnering' published by the Construction Industry Council (CIC). It is estimated that Option X12 is used on 9 per cent of NEC3 contracts.[4]

It must be stressed that no legal entity is created between the Partners, so it is not a partnership as such.

Some definitions need to be explained:

- 'The Partners' are those named in the Schedule of Partners.
- An 'Own Contract' is a contract between two Partners which includes Option X12.

- The 'Core Group' comprises the Partners listed in the Schedule of Core Group Members.
- 'Partnering Information' is information which specifies how the Partners work together.
- A 'Key Performance Indicator' is an aspect of performance for which a target is attached in the Schedule of Partners.

Each Partner, represented by a single individual, is required to work with the other Partners in accordance with the Partnering Information to achieve the Employer's objective stated in the Contract Data and the objectives of every other Partner.

The Core Group, led by the Employer's Representative, its members selected by the Partners, acts and takes decisions on behalf of the Partners. The Core Group also keeps up to date a Schedule of Core Group Members and a Schedule of Partners.

The Partners are required to work together, using common information systems, and a Partner may ask another Partner to provide information which he needs to carry out the work in his Own Contract. Each Partner gives an early warning to the other Partners when he becomes aware of anything that could affect another Partner's objectives.

The Core Group may give an instruction to the Partners to change the Partnering Information, which is a compensation event.

The Core Group also maintains a timetable showing the Partners' contributions. If the Contractor needs to change his programme to comply with the timetable, then it is a compensation event.

Each Partner also gives advice, information and opinion to the Core Group where required.

Each Partner must also notify the Core Group before subcontracting any work, though it does not say that the Core Group is required to respond to the notification.

Finally, Option X12 provides for Key Performance Indicators (KPIs) with amounts paid as stated in the Schedule of Partners. The Employer may add a KPI to the Schedule of Partners but cannot delete or reduce a payment.

Option X13: Performance bond

A performance bond is an arrangement whereby the performance of a contracting party (the Principal) is backed by a third party (the Surety), which could be a bank, insurance company or other financial institution, that should the Principal fail in his obligations under the contract, normally due to the Principal's insolvency, the Surety will pay a pre-agreed sum of money to the other contracting party (the Beneficiary). Whilst bonds are not always linked to insolvency, contracts normally give remedies in the event of default, eg Defects provisions, retention, withholding of payment, delay damages, etc. It is where the defaulter is not able to provide a remedy as he is insolvent that bonds normally provide the remedy.

This bond may be between any two contracting parties, Employer and Contractor, Employer and Nominated Subcontractor, Contractor and Domestic Subcontractor, or any other contractual relationship.

Note that as these are conditional bonds, it is normally a condition that default has to have occurred, there is no other means of remedy or recompense and the amount claimed is directly related to the default, before payment can be made to the Beneficiary.

If a performance bond is required from the Contractor it should be provided by the Contract Date. The amount of the bond, usually 10 per cent of the contract value, must be stated in Contract Data Part 1 and the form of the bond must be in the form stated in the Works Information. In the event of the default by the Contractor the Employer is not paid the full amount of the bond, but the amount of the bond is the maximum amount available to meet the Employer's costs incurred as a direct result of the default.

Normally, the value of the performance bond does not reduce, but they should have an expiry date, which could be Completion of the project, the Defects Date or may even include the 6-, 10- or 12-year limitation period following Completion of the Works to cover any liability for potential latent Defects, the expiry date being defined within the Works Information.

It is worth mentioning that payment of a performance bond is often very difficult and subject to a number of conditions, for example clear evidence of the Contractor's inability to perform his obligations, and also unequivocal evidence of the Employer's loss as a direct result of the Contractor's default, which in truth may not be fully known for some time after the Contractor's insolvency. Clearly, the expiry date for the bond has to be kept in mind if it takes some time to ascertain the Employer's actual loss as the right to claim may have expired.

A performance bond may be provided by a bank, an insurance company or a specialist bond provider. Clearly, if a bank provides a bond there may be some implications with regard to further credit as a safeguard should the bond be called, the bank also pursuing any amount paid from their customer, whereas with an insurance company the risk is accepted through the policy.

The bank or insurer which provides the performance bond must be accepted by the Project Manager.

It is important to note the difference between an insurance policy and a bond in that an insurance policy is a contract between *two* parties, the insured and the insurer, the insurer guaranteeing that he will pay a third party, who may not be specifically named, should an event occur, whilst a bond is a contract between *three* parties, the Contractor, the Employer and the Surety, all of whom are named within the agreement, no other party being able to benefit from the bond.

The cost of the performance bond will normally be governed by:

- the technical ability of the Contractor;
- the usual type of work undertaken by the Contractor;
- the size of the project;
- contracts already bonded for the Contractor;

- the overall management of the Contractor's business;
- the financial stability of the Contractor.

There are other types of bond available within construction contracts:

(I) ADVANCED PAYMENT BOND

This covers the Contractor's repayment of an advanced payment from the Employer to the Contractor. This is covered in the contract by Option X14.

(II) PAYMENT BOND

This covers the Employer's duty to pay the Contractor, or the Contractor's duty to pay his Subcontractors.

(III) BID OR TENDER BOND

Very often on particularly large international projects with a long lead in time, a bond may be required to ensure that the successful tenderer will not withdraw his tender and is able to proceed when required, in terms of his other commitments and his solvency.

In a similar vein, a scenario may be where a Contractor prices a tender based on a Subcontractor's Price and finds upon acceptance of his tender that the Subcontractor's offer is no longer open for acceptance, and the Contractor may find himself having to place orders with another Subcontractor for a higher Price.

It may be worthwhile in this event to bond the Subcontractor so that in the event that the acceptance period expires, the Contractor can recover monies through the bond.

(IV) RETENTION BONDS

Again, on particularly large projects it may be more practical to release retention monies early with a bond in place to protect the parties in the event of Defects arising.

(V) UNCONDITIONAL OR ON DEMAND BONDS

'Unconditional' or 'On Demand Bonds' can be redeemed by the Beneficiary, whether or not there has been a default within the contract and whether or not there has actually been a loss at all!

Most Employers will refrain from calling in bonds unnecessarily, but the entitlement remains should they wish to.

The 'knock-on' effect of calling in these bonds without due cause is that Contractors will price for that possibility within their tenders, as someone has to pay, and the result will be escalating Prices. Insurers are naturally very reluctant to accept on demand bonds.

CREATION OF BONDS

Bonds have to be made in writing and will be in the form of a deed, signed and sealed by the contracting parties.

The duration and financial limits of the Bond *must* be clarified and stated within the deed.

RELEASE AND CANCELLATION OF BONDS

The deed executed will normally specify the course of action to be taken in the event of failure of the Principal, leading to the calling in of the Bond by the Beneficiary.

Sometimes the Beneficiary has to give a form of notice to the Principal or he may have to sue for non-performance.

Only the Beneficiary can cancel the Bond, the Principal issuing a request to him, once he has 'performed'.

Option X14: *Advanced payment to the Contractor*

This is appropriate when the Contractor will incur significant 'up-front' costs before he starts receiving payments, for example in pre-ordering specialist Materials, Plant or Equipment. The payment is made within four weeks of the Contract Date or if an Advanced Payment Bond is required at the later of the Contract Date and the date the Employer receives the Bond. If the advanced payment is delayed, it is a compensation event.

If an Advanced Payment Bond is required, it is issued by a bank or insurer which the Project Manager has accepted, the Bond being in the amount that the Contractor has not repaid.

Advance Payment Bonds are a helpful security when an advance payment is made to a Contractor for Works to be performed; the Project Manager must accept the provider of that bond.

The amount of the payments is stated in the Contract Data Part 1 together with the repayment amounts and any requirements for a bond or security.

Where the advance payment reduces with time as, for example, stage payments are made against goods and/or services delivered under the contract, then the value of the Bond should reduce to reflect the outstanding amount of the advance payment.

Option X15: *Limitation of the Contractor's liability for his design to reasonable skill and care*

Whilst the Works Information defines what, if any, design is to be carried out by the Contractor, the contract is silent on the standard of care to be exercised by the Contractor when carrying out any design.

Two terms that relate to design liability are 'fitness for purpose' and 'reasonable skill and care'.

Whilst this book is intended for international use, in defining the term 'fitness for purpose' one must look in English law to the Sale of Goods Act 1979, which refers to the quality of goods supplied including their state and condition complying in terms of 'fitness for all the purposes for which goods of the kind in question are commonly supplied'.

In construction, fitness for purpose means producing a finished project fit in all respects for its intended purpose. This is an absolute duty independent of negligence, and in the absence of any express terms within the contract to the contrary, a Contractor who has a design responsibility will be required to design and build the project 'fit for purpose'.

Some contracts will limit the Contractor's liability to that of a consultant, ie reasonable skill and care. The ECC does this through Option X15. If this Secondary Option is not chosen, the Contractor's liability for design is 'fitness for purpose'.

Contrast this with the level of liability of a consultant (whether acting for an Employer or a Contractor) in providing a design service, which is defined, again in English law, by the Supply of Goods and Services Act 1982, where there is an implied term that the consultants will carry out the service, in this case design, with reasonable care and skill, which means designing to the level of an ordinary, but competent person exercising a particular skill.

So, in the absence of any express terms to the contrary, a Designer will normally be required to design using 'reasonable skill and care'. This is normally achieved by the Designer following accepted practice and complying with national standards, codes of practice, etc.

Clearly, despite the Contractor believing that he has offset his design obligations and liability to his designing consultant, he has to be aware that he and his consultant have differing levels of care and liability.

Option X16: Retention (used only with Options A, B, C, D and E)

The Employer may retain a proportion of the Price for Work Done to Date once it has reached any retention free amount, the retention percentage and any retention free amount being stated in Contract Data Part 1.

Under Clause 11.2(23) (Options C, D, E and F), Defined Cost does not include amounts deducted for retention, so, if the Contractor deducts retention from a Subcontractor, the figure before the deduction is used in calculating Defined Cost, ensuring that there is no 'double deduction' of retention.

Following Completion of the whole of the Works, or the date the Employer takes over the whole of the Works, whichever happens earlier, the retention percentage is halved, and then the final release is upon the issue of the Defects Certificate.

It is important to note that retention is held against undiscovered Defects and not incomplete work.

Option X17: Low performance damages

In the event that the Contractor produces defective work, the Employer has three options:

1. The Contractor corrects the Defect (Clause 43.1).
2. If the Contractor does not correct the Defect, the Project Manager assesses the cost to the Employer of having the Defect corrected by other people and the Contractor pays this amount (Clause 45.1).
3. The Employer can accept the Defect and a quotation from the Contractor for reduced Prices and/or an earlier Completion Date (Clause 44).

Where the performance in use fails to reach the specified level within the contract, the Employer can take action against the Contractor to recover any damages suffered as a result of the breach, but as an alternative can recover low performance damages under Option X17 if it has been selected.

Example

The Works Information requires the Contractor to design and install a HVAC system to a major retail development. There are specific and measurable performance criteria for the system including temperature variations and energy efficiency. The system is required to have a design life of 30 years (360 months).

The Works Information states that the system will be tested at Completion and includes a table showing how the performance of the system will be measured and acceptable levels of achievement.

The system is expected to perform to 98–100% of performance criteria. If it falls within 90–98% the system will be accepted, but low performance damages will be payable by the Contractor to the Employer. If it falls below 90% it will not be accepted.

Contract Data Part 1 contains the following entries:

Amount per month	Performance level
£100.00	96–98%
£150.00	94–96%
£200.00	92–94%
£250.00	90–92%

When tested, the system achieves 95% of the performance criteria.

Therefore, 360 months × £150.00 = £54,000 is payable by the Contractor at the Defects Date. Contrary to popular belief, the Employer does not collect the damages from the Contractor every month!

Option X18: Limitation of liability

If the Contractor causes any loss of or damage to the Employer's property, he would normally be liable for the full cost of remedial work, however Clause X18 provides for this liability to be limited to amounts stated in the Contract Data.

In addition, the Contractor's liability to the Employer for latent Defects due to his design may again be limited to amounts stated in the Contract Data.

Clause X18.4 can be used to place limits on the total liability the Contractor has to the Employer for all matters under the contract other than excluded matters in contract, tort or delict.

Excluded matters are:

- loss or damage to Employer's property;
- delay damages if Option X7 applies;
- low performance damages if Option X17 applies;
- Contractor's share if Option C or D applies.

The Contractor is not liable for any matter unless it has been notified to the Contractor before the end of liability date which is stated in the Contract Data in terms of years after the Completion of the whole of the Works. In the UK, this may be 6 or 12 years dependent on the type of contract, other legislations set this at 10 years.

Option X20: Key Performance Indicators

Performance of the Contractor can be monitored and measured against Key Performance Indicators (KPIs) using Option X20. Targets may be stated for Key Performance Indicators in the Incentive Schedule.

The Contractor is required to report his performance against KPIs to the Project Manager at intervals stated in the Contract Data including the forecast final measurement. If the forecast final measurement will not achieve the target stated in the Incentive Schedule the Contractor is required to submit his proposals to the Project Manager for improving performance.

The Contractor is paid the amount stated in the Incentive Schedule if the target for a KPI is improved upon or achieved. Note that there is no payment due from the Contractor if he fails to achieve a stated target.

The Employer may add a new KPI and associated payment to the Incentive Schedule but may not delete or reduce a payment.

Option Y(UK)2: The Housing Grants, Construction and Regeneration Act 1996

This Option is only applicable to UK contracts where the Housing Grants, Construction and Regeneration Act 1996 applies and deals only with the payment aspects of the Act. Adjudication under the Act is covered by Option W2.

Whilst this book avoids discussing NEC3 specifically in connection with UK construction law, it is worth mentioning the Local Democracy, Economic

Development and Construction (LDEDC) Act 2009 which has a direct bearing on Option Y(UK)2.

The LDEDC Act, specifically Part 8, amends Part II of the Housing Grants, Construction and Regeneration Act 1996, and came into force in England and Wales on 1 October 2011 and in Scotland on 1 November 2011.

It is important to note that as the new Act amends the Housing Grants, Construction and Regeneration (HGCR) Act 1996, you have to take into account both Acts to fully understand how it applies to your contract.

The Act applies to all construction contracts including:

- construction, alteration, repair, maintenance, etc;
- all normal building and civil engineering work, including elements such as temporary works, scaffolding, Site clearance, painting and decorating;
- consultants' agreements concerning construction operations;
- labour-only contracts;
- contracts of any value.

It excludes:

- extraction of oil, gas or minerals;
- supply and fix of Plant in process industries, eg nuclear processing, power generation, water or effluent treatment;
- contracts with residential occupiers;
- header agreements in connection with PFI contracts;
- finance agreements.

The HGCR Act 1996 only applied to contracts in writing, but the LDEDC Act 2009 now also applies to oral contracts.

Briefly, the differences between the HGCR Act 1996 and the LDEDC Act 2009 include:

- A notice is to be issued by the payer to the payee within five days of the payment due date, or by the payee within not later than five days after the payment due date, the amount stated is the notified sum. The absence of a notice means that the payee's application for payment serves as the notice of payment due and the payer will be obliged to pay that amount.
- A notice can be issued at a prescribed period before the final date for payment which reduces the notified sum.
- The Contractor has the right to suspend all or part of the Works, and can claim for reasonable costs in respect of costs and expenses incurred as a result of this suspension as a compensation event.
- Terms in contracts such as 'the fees and expenses of the Adjudicator as well as the reasonable expenses of the other party shall be the responsibility of the party making the reference to the Adjudicator' have been prohibited. Much has been written about the effectiveness of such clauses, and whether

they comply with the previous Act, so the new Act should provide clarity for the future.

- The Adjudicator is permitted to correct his decision so as to remove a clerical or typographical error arising by accident or omission. Previously he could not make this correction.

NB The publishers have issued a brief amendment to the ECC to align it with the Local Democracy, Economic Development and Construction Act 2009.

Option Y(UK)3: The Contracts (Rights of Third Parties) Act 1999

This Option is only applicable to UK contracts where the Contracts (Rights of Third Parties) Act 1999 applies.

The principle of privity of contract is that a person who is not a party to a contract cannot enforce the terms of that contract.

The Contracts (Rights of Third Parties) Act 1999 effectively abolishes privity of contract as an absolute principle, enabling a person who is not a party to a contract, a 'third party', to enforce a term of the contract if the contract expressly provides that he may. The third party must be expressly identified in the contract, either by name or as a member of a class or by a particular description, but does not need to be in existence at the date of the contract.

Under the Act, there is available to the third party any remedy that would have been available to him in an action for breach of contract as if he had been a party to the contract.

The parties to the contract are prohibited from making any agreement to vary the contract in such a way as to adversely affect the third party's entitlement to enforce the contract without the third party's consent.

The Act names two specific parties:

1. 'the promisor' means the party to the contract against whom the term is enforceable by the third party; and
2. 'the promisee' means the party to the contract by whom the term is enforceable against the promisor.

Whilst the Act is not specially aimed at the construction industry, the industry spends a lot of time and money in constructing collateral warranties (see Chapter 2) enabling a multitude of parties to have certain rights over specified periods of time, so it is now possible to draft building contract documents to confer benefits upon the parties who would have had the benefit of collateral warranties.

The construction industry has generally not been in favour of this Act, preferring to use collateral warranties, but it must be said that the Act has several advantages in terms of time and cost in that only one document needs to be drafted, though it is fair to say that many parties prefer a warranty as specific evidence of their rights. Banks and funders in particular like the comfort of a good old-fashioned document.

Option Z: Additional conditions of contract

Option Z allows conditions to be added to or omitted from the core clauses.

All changes to the core clauses should be included as Z clauses rather than amending the core clauses themselves, so in effect the clause remains in the contract, but is amended within the Z clause. It is also critical that when drafting a Z clause, it must be clearly stated what happens to the original core clause, for example it is deleted. If the original core clause is not deleted, then it is likely than an inconsistency will arise which will be interpreted against the party who wrote or amended the clause, ie the Employer.

These conditions may modify or add to the core clause, to suit any risk allocation or other special requirements of the particular contract. However, changes should be kept to a minimum, consistent with the objective of using industry standard, impartially written contracts.

It must be remembered that if an Employer amends a contract to allocate a risk to the Contractor which may have been intended to be held by the Employer, the Contractor has a right to price it in terms of time and money, therefore the practice of Employers amending contracts to pass risk to Contractors without considering who is best able to price, control and manage those risks can in many cases prove to be unwise and uneconomical.

It is important to spend time considering whether a Z clause is appropriate in each case, then when that decision has been made, that the clause is drafted correctly and aligned to the drafting principles of the original contract; in the case of the ECC, use ordinary language, present tense, short sentences and bullet pointing.

It is not unusual to see Z clauses in an ECC contract written in a legalistic language, in the future tense and without punctuation apart from full stops!

Some correctly drafted examples

Payment

Assessing the amount due

Delete Clause 50.4

Insert:

The *Contractor* submits an application for payment one week before each *assessment date*. In assessing the amount due, the *Project Manager* considers the application for payment the *Contractor* submits. The *Project Manager* gives the *Contractor* details of how the amount due has been assessed.

The Contractor's share

Delete Clause 53.3

Insert:

The *Project Manager* makes a preliminary assessment of the *Contractor's* share at any assessment date. If the forecast Price for Work Done to Date is less than the final Total of the Prices, this share is included in the amount due following Completion of the whole of the Works. If the forecast Price for Work Done to Date is more than the forecast Total of the Prices, this share is retained from the amount due.

For Options C, D, E & F

Delete Clause 11.2(29)

Insert:

The Price for Work Done to Date is the total Defined Cost which the Contractor has paid plus the Fee.

For Options C, D and E

Clause 11.2(29), sixth bullet

Delete the words 'after Completion'

0.5 The roles of the parties

There are a number of parties referred to in the ECC:

- the Employer;
- the Contractor;
- the Project Manager;
- the Supervisor;
- the Subcontractor;
- the Adjudicator.

The Employer

The Employer is the Client, and the contract is between him and the Contractor, though the Employer is rarely mentioned in the contract.

The Employer's obligations are:

- to act as stated in the contract and in a spirit of mutual trust and co-operation;
- to allow the Contractor access to and use of each part of the Site for the Contractor to carry out the work included in the contract. This access must be allowed on or before the later of the access date (the date in the Contract Data) and the date for access shown on the Accepted Programme;
- to allow the Contractor access to and use of a part of the Works which he has taken over if they are needed for correcting a Defect (subject to Clause 45.2);

- to make payment to the Contractor in accordance with the Project Manager's certificates;
- to take over the Works at Completion.

The Contractor

The Contractor and the Employer are the parties to the contract, the Contractor's responsibility being to Provide the Works in accordance with the contract, the Works Information describing the Works and any constraints on how he has to provide it.

There are also obligations in respect of complying with instructions, notifying early warnings, submitting programmes for acceptance, and notifying and submitting quotations for compensation events which are highlighted in later chapters.

The Project Manager

The Project Manager is appointed by the Employer and manages the contract on his behalf (see Appendix to this Introduction for list of Project Manager actions).

Other forms of contract tend to name an Engineer or Architect, who acts for the Employer, and in addition to having design responsibilities, he also administers the contract. The Engineering and Construction Contract separates the role of Designer from that of managing the project and administering the contract.

The Project Manager is named in Contract Data Part 1 and may be appointed from the Employer's own staff or may be an external consultant. If a consultant, they may be appointed using the NEC3 Professional Services Contract. The Project Manager is a named individual not a company.

It is vital that the Employer appoints a Project Manager who has the necessary knowledge, skills and experience to carry out the role, which includes acceptance of designs and programmes, issuing early warnings, certifying payments and dealing with compensation events. Disciplines which have carried out the Project Manager role to date include Engineers from a civil engineering, structural or process background, Architects, Building Surveyors and Quantity Surveyors, in addition to Project Managers themselves.

It is also vital that the Employer gives the Project Manager full authority to act for him, particularly when considering the time scales imposed by the contract. If the Project Manager has to seek approvals and consents from the Employer, then he must do so, and the appropriate authority must be given, in compliance with the contractual time scales. If the approval process is likely to be lengthy, it is critical that both the Employer and his Project Manager set up an accelerated process to comply with the contract or that the time scales in the contract are amended. The former is the vastly preferred method in order that NEC3 will work to its full effect.

Although the Project Manager manages the contract at the post-contract stage, he is usually appointed pre-contract to deal with matters such as feasibility issues, advising on design, procurement, cost-planning tendering and programme matters. Note that, unlike many other contracts, the ECC does not name the Quantity Surveyor, so this would need to be considered in dealing with financial matters pre- and post-contract. Either the Project Manager may carry out the role of the Quantity Surveyor himself or delegate to another.

The role of a Project Manager within the ECC is different to the usual concept of a Project Manager and it is important that this difference is recognised as it is a traditional area of misunderstanding amongst users of the contract.

There should only be one Project Manager for the project, though the Project Manager can delegate responsibilities to others after notifying the Contractor, and may subsequently cancel that delegation (Clause 14.2).

As with any other notification under the contract, if the Project Manager (or the Supervisor) wishes to delegate, the notice to the Contractor must be in a form which can be read, copied and recorded, ie by letter or email, not by verbal communication such as a telephone call.

The delegation may be due to the Project Manager being absent for a period due to holidays or illness, or because he wishes to appoint someone to assist him in his duties.

Although the contract is not specific, the notice should identify who the Project Manager is delegating to, what their authority is, and how long the delegation will last, so the Contractor is in no doubt as to who has authority under the contract.

It is important to recognise that in delegating the Project Manager is sharing an authority with the delegate who will then represent him, he is not passing responsibility on to them. The authority is shared, but ultimately responsibility will remain with the Project Manager as the principal.

It is also important to note that one can only delegate outwards from the principal, ie no one can assume authority possibly because, within an organisation, they are senior to the Project Manager.

If the Employer wishes to replace the Project Manager or the Supervisor he must first notify the Contractor of the name of the replacement before doing so.

Designers

Designers are not named parties in the ECC as the design may be carried out on behalf of the Employer and/or the Contractor. The Employer is required to appoint Designers for his own design, and the Contractor for his design. The concept of novation of Designers is considered in Chapter 2. It is important to recognise that the Designer does not have the same authority as, say, an Architect under the JCT contracts, who has the dual role of Designer and Contract Administrator. Under the ECC the Designer only designs, he does not administer the contract, and therefore any issuing of drawings, other design information and associated instructions must be done through the Project Manager. Again, the NEC3 Professional Services Contract may be used for appointment of Designers, though its use is not mandatory.

The Supervisor

The Supervisor is appointed by the Employer and manages issues regarding testing and Defects on his behalf. (See Appendix to this Introduction for list of Supervisor actions.)

Again, as with the Project Manager, the Supervisor is named in Contract Data Part 1 and may be appointed from the Employer's own staff or may be an external consultant. Again, as with the Project Manager, the Supervisor is a named individual not a company. It is possible that, provided he has sufficient expertise and time available, the Supervisor and the Project Manager could be the same person. It is not unusual for the Employer's Designer and the Supervisor to be the same person as he is well qualified, having done the design, to inspect the Works to make sure the Contractor has complied with it.

The title Supervisor is not normally found in other contracts, many seeing the role as similar to that of a Clerk of Works or Resident Engineer, and in many ways the role is similar, but it is important to recognise that those roles are usually delegated by others within their respective contracts, therefore giving the Clerk of Works or Resident Engineer little or no ultimate responsibility.

Some practitioners have commented that the term 'Supervisor' sounds as if he is managing the Contractor; however, it is important to recognise that the Contractor is required to take a proactive role in carrying out the Works, and, correcting his own Defects, the Supervisor manages this process on behalf of the Employer.

Disciplines which have carried out the Supervisor role to date include Engineers from a civil engineering, structural or process background, Architects and other design disciplines, Building Surveyors and Clerks of Works.

It is important to recognise that the Supervisor does not act on behalf of the Project Manager; he represents the Employer and has his own responsibilities and obligations within the contract which may be summarised as follows:

- The Supervisor carries out and/or witnesses tests being carried out by the Contractor or some other party.
- The Supervisor may instruct the Contractor to search for a Defect.

   This is common to other contracts where it is referred to as 'opening up' or 'uncovering', the principle being if no Defect is found the matter is dealt with as a compensation event.

- The Supervisor notifies the Contractor of each Defect as soon as he finds it.
- The Supervisor issues the Defects Certificate at the later of the Defects date and the end of the last Defect Correction Period.

   This is a very important action as once the Defects Certificate has been issued, the Contractor is no longer required to provide the insurances under the contract. Also, dependent on which Secondary Options have been selected, any retention (Clause X16) is released back to the Contractor, if a Defect included in the Defects Certificate shows low performance, the Contractor pays low performance damages (Clause X17), and the Contractor no longer has to report performance against KPIs (Clause X20). The Defects Certificate is issued on the later of the Defects Date and the last Defect Correction Period, not when all Defects have been corrected. In this case, the Defects Certificate may show Defects which the Contractor has not corrected.

- The Supervisor marks Plant and Materials as for the contract if the contract identifies them for payment.

  Marking Plant and Materials can involve making a physical mark on them to denote that he has seen them but can include compiling an inventory, photographic records, etc. The Contractor is required to prepare the Plant and Materials for marking as required by the Works Information, which can include setting them aside from other stock, protection, insurances and any vesting requirements. Once they have been marked, any title the Contractor has to them passes to the Employer.

Again, as with the Project Manager, there should only be one Supervisor, who may then delegate duties and authorities to others. On a large project it is common for various personnel to be responsible for checking mechanical, electrical, structural, finishings, landscape elements, etc, but again there should be one Supervisor who delegates duties to others in respect of these elements. Each then reports back to the single Supervisor.

The Adjudicator

The Adjudicator can be chosen by one of three methods:

1. By the Employer specifically naming the individual in Contract Data Part 1. The Contractor therefore knows at the time of tender who the Employer has named and can therefore confirm if he does not agree with the Employer's choice. This has advantages in that the parties always know who will be the Adjudicator if they have a dispute. The disadvantages are first that parties have commented in the past that naming the individual within the Contract Data assumes that there will be a dispute, and often the Adjudicator will require payment for having his name in the contract and being available should the parties have that dispute!
2. By the Employer naming an Adjudicator Nominating Body in Contract Data Part 1. This would normally be a professional institution. This tends to be the favoured method.
3. By the parties agreeing who will be the Adjudicator at the time of the dispute. The disadvantage of this method is that once the parties are in dispute they probably would not readily agree with each other on anything!

The Adjudicator becomes involved only when either contracting party refers a dispute to him. As an impartial person he is required to give a decision on the dispute within stated time limits. If either party does not accept his decision they may refer the dispute to the tribunal (normally litigation or arbitration). The Adjudicator's Contract requires that payment of the Adjudicator's fees is shared by the parties unless otherwise stated.

0.6 Mutual trust and co-operation

The first clause of all the NEC3 contracts requires the parties to act in a spirit of mutual trust and co-operation, the ECC requiring the Employer, the Contractor, the Project Manager and the Supervisor to do so.

This mirrors Sir Michael Latham in his report *Constructing the Team* when he recommended that the most effective form of contract should include 'a specific duty for all parties to deal fairly with each other and in an atmosphere of mutual co-operation'.

It is worth considering this obligation in a little detail, as the clause has often been viewed with some confusion, and for those who have spent many years in the construction industry, with a great degree of scepticism.

The first part 'the Employer, the Contractor, the Project Manager and the Supervisor shall acted as stated in the contract' is straightforward, and many would say is not required to be expressly stated as the named parties have certain obligations within the contract even though the Project Manager and the Supervisor are not the contracting parties, but the second part 'and in a spirit of mutual trust and co-operation' has caused some debate.

Most practitioners state that their understanding of the second part of this clause is that the parties should be non-adversarial towards each other, acting in a collaborative way and working for each other rather than against each other, and in reality that is what the clause requires, and in reality is how all parties to all contracts should behave.

However, the difficulty is that if a party does not act in a spirit of mutual trust and co-operation, what can another party who may be affected do? The answer is that the clause is almost unenforceable as it is virtually impossible to define and quantify the breach, or the ensuing damages that flow from the breach. In addition, the Contractor is not contractually related to the Project Manager or Supervisor, so either would be unable to take direct action against the other for breach of contract other than through the Employer, who would ultimately be liable for his Project Manager and/or Supervisor complying with the contract.

To that end, Employers have been seen to insert a Z clause to delete Clause 10.1; however it is strongly recommended that it should remain in the contract, if merely viewed as a statement of good intent. Cynics may say if you delete the clause the parties are not required to act in a spirit of mutual trust and co-operation, but that view can only remain in the domain of cynics!

In effect, a clause requiring parties to act in a certain spirit will probably not, on its own, have any real effect. Within the ECC, it is the clauses that follow within the contract which require early warnings, clearly detailed programmes which are submitted for acceptance, and a structured change management process, that actually create and develop that level of mutual trust and co-operation rather than simply inserting a statement within the contract requiring the parties to do so.

0.7 Some unique NEC3 terms

There are some terms which, whilst not being used in other contracts have different meanings within NEC3 contracts.

(i) Equipment and Plant

The ECC definitions of Equipment and Plant are different to many other contracts.

Equipment: Clause 11.2(7)

Equipment is defined in the contract as 'items provided by the Contractor and used by him to Provide the Works and which the Works Information does not require him to include in the Works'. In that sense it includes anything which is used to build the project, but which is not incorporated into the Works and does not remain on Site after Completion. Specific examples of Equipment would include excavators, dumpers, lorries, cranes, tools, scaffolding and temporary accommodation. These items are normally referred to as 'Plant' in other contracts.

It also includes associated items such as fuel, lubricants, tyres and other consumables.

Plant: Clause 11.2(12)

Plant (and Materials) are 'items intended to be included in the Works', specific examples being mechanical and electrical installations.

It is critical that practitioners recognise the difference between Equipment and Plant and continue to use the correct terms as they are defined separately throughout the contract, and particularly within the Schedule of Cost Components and the Shorter Schedule of Cost Components when considering payments and compensation events.

(ii) Others

The ECC uses the term 'Others' to define parties who are 'not the Employer, the Project Manager, the Supervisor, the Adjudicator, the Contractor or any employee, Subcontractor or Supplier of the Contractor'. Essentially Others are anyone outside the contract, for example other Contractors, statutory bodies, regulatory authorities, utilities companies, etc.

References to Others include the Contractor's obligation to co-operate with Others in obtaining and providing information, co-operating with Others and sharing the Working Areas, obtaining approval of his design from Others where necessary, the Contractor providing access to Plant and Materials being stored for Others notified by the Project Manager (Contractor's Obligations), the order and timing of the work of Others (Time), loss or damage to Plant and Materials supplied by Others (Risk and insurance), the Contractor substantially hindering Others (Termination).

(iii) *Working Areas: Clause 11.2(18)*

The term 'Working Areas' needs a little explanation as some confusion has arisen amongst ECC users.

The contract defines the Working Areas under Clause 11.2(18) as those parts of the Working Areas which are:

- necessary for Providing the Works; and
- used only for work in this contract.

The Working Areas are initially the Site, the boundaries of which are defined in Contract Data Part 1, but also any additional Working Areas may be identified by the Contractor in Contract Data Part 2 and submitted as part of his tender.

An example of an addition to the Working Areas could be where the footprint of the project to be built fills the Site, in which case the Contractor may name an additional area such as an adjacent field which he proposes to use for storage or to locate his Site compound. In doing so, in order to qualify as an addition to the Working Areas, the Contractor should comply with Clause 11.2(18) in that this additional area is necessary for Providing the Works and is used only for work in this contract.

The Contractor may also submit a proposal to the Project Manager for adding to the Working Areas during the carrying out of the Works. The Project Manager may refuse acceptance because the Contractor has not complied with Clause 11.2(18).

The implications of adding to the Working Areas depend on the Main Option chosen.

Options A and B

The Shorter Schedule of Cost Components refers to the cost of resources used within the Working Areas, so these resources if also used within the extended Working Areas would be included as cost rather than within the Fee percentage when assessing compensation events.

Options C, D and E

The Schedule of Cost Components refers to the cost of resources used within the Working Areas, so these resources if also used within the extended Working Areas would be included as cost rather than within the Fee percentage when assessing compensation events.

(iv) Communications: Clause 13

NEC3 has strict rules regarding communications under the contract:

- Each instruction, certificate, submission, proposal, record, acceptance, notification, reply and other communications must be communicated in a form which can be read, copied and recorded. The communication has effect when it is received. This may be by electronic means, for example email, or via a project intranet, so the issue and receipt are simultaneous. Normally the parties agree a protocol for project communications.
- This requirement is particularly important in respect of instructions as the contract does not recognise verbal instructions and whereas under other forms of contract a verbal instruction can be confirmed by the Contractor and if not dissented from, normally within 7 or 14 days it becomes an instruction, the NEC3 contracts do not have that provision. The instruction may be in the form of a letter, a pro forma instruction, or an email.
- The Project Manager, Supervisor or Contractor replies within the period for reply stated in Contract Data Part 1, unless the contract states otherwise. This period can be extended by mutual agreement between the communicating parties.
- If the Project Manager is required to accept or not accept, he must state his reasons for non-acceptance. Withholding acceptance for a reason stated in the contract is not a compensation event; this is particularly relevant in respect of:

  - acceptance of Contractor's design;
  - acceptance of a proposed Subcontractor;
  - acceptance of a Contractor's programme.

- The Project Manager issues his certificates (Payment, Completion and Termination Certificates) to the Employer and the Contractor, the Supervisor issues the Defects Certificate to the Project Manager and the Contractor.
- Notifications should be communicated separately from other communications, so for example, an early warning notice should not be included as part of a set of meeting minutes, or a letter which includes other subjects.

0.8 Subcontracts

The ECC has been designed on the assumption that work may be subcontracted with Clause 26 dealing specifically with the appointment of Subcontractors. Note also, the ECC definition of a Subcontractor under Clause 11.2(7):

A Subcontractor is a person or organisation who has a contract with the Contractor to:

- construct or install part of the Works;
- provide a service necessary to Provide the Works;

- supply Plant and Materials which the person or organisation has wholly or partly designed specifically for the Works.

This is quite a wide definition, and particularly the third bullet point which includes a person or organisation which may be a Supplier of Plant or Materials, but has a design responsibility. This would include companies who, for example, supply roof trusses, precast concrete bridge or floor units, joinery, or even purpose-made bricks because they have a design responsibility, even though they may have no responsibility for fixing the items. In many contracts companies without responsibility for fixing what they supply would be classed as Suppliers.

If the Contractor proposes to subcontract work he is responsible for Providing the Works as if he had not, and also that the contract applies as if a Subcontractor's employees and Equipment were his own, thus emphasising that the Contractor is wholly liable to the Employer for any Subcontractors.

Clause 26.2 is worth specific mention as the Contractor is obliged to submit the name of each proposed Subcontractor to the Project Manager for acceptance. No Subcontractor can be appointed until the Project Manager has given his acceptance. Note again that the submission by the Contractor and the acceptance by the Project Manager must both be in a form which can be read, copied and recorded.

As the Contractor cannot appoint the particular Subcontractor until the Project Manager has accepted him, it is vital that the Contractor submits the name of his Subcontractors as early as possible so that the Project Manager has a reasonable period in which to consider them. The Project Manager must accept, or give his reasons for not accepting, the Subcontractor within the 'period for reply' stated in the contract.

A reason for the Project Manager not accepting the Subcontractor is that 'his appointment will not allow the Contractor to Provide the Works'. This clause probably needs a word of caution to Project Managers who may give the Contractor any one of several reasons for not accepting the Subcontractor, but if the reason is not that 'his appointment will not allow the Contractor to Provide the Works', then it is a compensation event under Clause 60.1(9), 'the Project Manager withholds acceptance for a reason not stated in this contract'.

Examples of non-acceptance of a Subcontractor which could give rise to a compensation event are 'I have never heard of your proposed Subcontractor', or 'I have heard that he is not doing too well on one of your other Sites', neither of which comply with the reason 'his appointment will not allow the Contractor to Provide the Works'.

Whilst the contract does not state as such, clearly, if the Project Manager has concerns about a Subcontractor he should raise them with the Contractor who should be expected to provide reasonable substantiation of their competence, possibly in the form of examples of previous work and references. If, despite the Contractor providing this substantiation, the Project Manager refuses to accept the Subcontractor, then it is a compensation event.

Project Managers are often concerned about accepting a particular Subcontractor; probably the best advice is to ask the Contractor for some substantiation and unless the substantiation proves the Project Manager's concerns, to accept the Subcontractor and remind the Contractor that under Clause 26.1 responsibility for that Subcontractor remains with the Contractor.

Whilst the contract does not require the Contractor to use an NEC contract for his Subcontractors, under Clause 26.3 he is required to submit the proposed conditions of contract for his Subcontractors to the Project Manager for acceptance unless an NEC contract is used, or the Project Manager has agreed that no submission is required.

As the Contractor cannot appoint the particular Subcontractor on the proposed subcontract conditions until the Project Manager has accepted them, the Contractor should submit proposed conditions of contract as early as possible so that the Project Manager has a reasonable period in which to consider them.

A reason that the Project Manager can give for not accepting the contract conditions is that:

- they will not allow the Contractor to Provide the Works; or
- they do not include a statement that the parties to the subcontract shall act in a spirit of mutual trust and co-operation.

In addition, Options C, D, E and F also require the Contractor to submit the proposed Contract Data for each subcontract to the Project Manager for acceptance. This is an important requirement as costs arising from the subcontracts will become due for payment as Defined Cost.

The NEC3 Engineering and Construction Subcontract is identical to the ECC, providing back to back provisions, but including a number of changes appropriate to subcontracts.

General

- There is no Project Manager or Supervisor, so all actions are between the Contractor and the Subcontractor.
- Option F does not exist in the subcontract, this is because under Option F all the Works are subcontracted by the Contractor, so a management contract would not be applicable between a Contractor and a Subcontractor.
- 'Contract Date' and 'Completion Date' are replaced by 'Subcontract Date' and 'Subcontract Completion Date'.
- 'Works Information' is replaced by 'Subcontract Works Information', though the term 'Site Information' is not changed as it still relates to 'the Site'.

The Subcontractor's main responsibilities

- Clause 26 related to 'Subcontracting' now refers to 'Subsubcontracting'.

Payment

- The Contractor certifies a payment within two weeks of each assessment date; under the ECC, the Project Manager certified each payment within one week of each assessment date.
- Each certified payment is made within four weeks of each assessment date.

Compensation events

- The compensation events are the same excepting that the reference to the Employer, the Project Manager and the Supervisor in the ECC is replaced by the Contractor, though Clauses 60.1(5), (14) and (16) include the Employer, and Clause 60.1(11) refers to a test or inspection done by the Contractor or the Supervisor causing unnecessary delay.
- The Subcontractor submits quotations within one week of being instructed to do so by the Contractor, the Contractor replies within four weeks.

Risks and insurance

- Clause 80.1 lists the Employer's and the Contractor's risks.

Where the subcontract Works comprise straightforward work, and imposes low risks on the Contractor and the Subcontractor, the Engineering and Construction Short Subcontract may be used. Again, the NEC3 Engineering and Construction Short Subcontract is identical to the Engineering and Construction Short Contract, but includes a number of changes appropriate to subcontracts.

Nominated Subcontractors

Although many other contracts provide for Nominated Subcontractors chosen by the Employer, or his representative, and the Contractor is instructed to use them, the NEC contracts have never done so.

This is for a number of reasons:

- The Contractor should be responsible for managing all that he has contracted to do. Nomination will often split responsibilities.
- The use of Prime Cost Sums in bills of quantities means that the Contractor does not have to price that element of the work other than for profit and attendances, the more Prime Cost Sums, the less the tenderers have to price and thus there is less pricing competition between the tenderers.
- Contracts will often give the Contractor relief in the form of an extension of time in the event of a default by the Nominated Subcontractor, provided that Contractor has done all that would be reasonable or practicable to manage the default.
- Contracts will often give the Contractor relief in the form of loss and expense

where it cannot be recovered from the Nominated Subcontractor, again provided the Contractor had done all that would be reasonable or practicable to manage the default. This would include the insolvency of a Nominated Subcontractor and a subsequent renomination.

0.9 Appendices to Introduction

Project Manager actions

The Project Manager:

General:

10.1 acts as stated in the contract and in a spirit of mutual trust and co-operation
13.3 replies to a communication within the period for reply
13.5 may extend the period for reply if the Contractor agrees
13.6 issues his certificates to the Employer and Contractor
13.8 may withhold acceptance of a submission by the Contractor
14.2 may delegate any of his actions, after notifying the Contractor
14.3 may give an instruction which changes the Works Information or a Key Date
16.1 gives an early warning by notifying the Contractor
16.2 may instruct the Contractor to attend a risk reduction meeting
16.4 revises the Risk Register at each risk reduction meeting
17.1 notifies the Contractor as soon as aware of an ambiguity or inconsistency
19.1 gives an instruction to the Contractor stating how he is to deal with the prevention event.

Contractor's main responsibilities:

21.2 accepts or does not accept the Contractor's design
23.1 accepts or does not accept the Contractor's design of Equipment
24.1 accepts or does not accept the Contractor's replacement person
24.2 may instruct the Contractor to remove an employee
25.3 assesses the additional cost incurred by the Employer for Contractor failing to meet a Key Date
26.2 accepts or does not accept the Contractor's proposed Subcontractor
26.3 accepts or does not accept the proposed conditions of subcontract.

Time:

30.2 decides the date of Completion
31.1 receives the Contractor's first programme
31.3 accepts or does not accept the Contractor's programme
32.2 accepts or does not accept the Contractor's revised programme
34.1 may instruct the Contractor to stop or not to start any work

35.3 certifies the date upon which the Employer takes over a part of the Works

36.1 instructs the Contractor to submit a quotation for acceleration.

Testing and Defects:

40.6 assesses the cost to the Employer of repeating a test after a Defect is found

43.4 arranges for the Employer to allow the Contractor access to any part of the Works

44.1 may accept a proposal from the Contractor to change the Works Information to accept a Defect

44.2 accepts or does not accept the Contractor's quotation for accepting a Defect

45.1 assesses the cost of having a Defect corrected by other people

45.2 assesses the cost to the Contractor of correcting the Defect.

Payment:

50.1 assesses the amount due at each assessment date

50.4 considers any application for payment from the Contractor

50.5 corrects any wrongly assessed amount due

51.1 certifies a payment within one week of each assessment date.

Compensation events:

61.1 notifies the Contractor of a compensation event

61.2 may instruct the Contractor to submit a quotation for a proposed instruction

61.3 is notified by the Contractor of a compensation event

61.4 notifies the Contractor of his decision that the Prices, the Completion Date and the Key Dates are not to be changed

61.5 notifies the Contractor that he did not give an early warning

61.6 states assumptions about an event too uncertain to be forecast reasonably, and notifies a correction to an assumption

62.1 may instruct the Contractor to submit alternative quotations

62.3 replies within two weeks of receipt of a Contractor's quotation

62.4 instructs the Contractor to submit a revised quotation

62.5 extends the time allowed for submitting or replying to a quotation

62.6 replies to the Contractor's notification

63.5 assesses a compensation event as if the Contractor had given early warning

63.9 corrects the description of the Condition for a Key Date

64.1 assesses a compensation event

64.2 assesses a compensation event using his own assessment of the programme for remaining work

64.3 notifies the Contractor of his assessment of a compensation event

64.4 replies to the Contractor's notification

65.1 implements each compensation event.

Title:

70.2 permits the Contractor to remove Plant and Materials
72.1 allows the Contractor to leave Equipment in the Works
73.1 instructs the Contractor in respect of objects of value or historical interest.

Risks and insurance:

85.1 receives the Contractor's insurance policies and certificates
87.1 submits the Employer's insurance policies and certificates to the Contractor.

Termination:

90.1 is notified of a reason for termination and issues Termination Certificate
91.2 and 91.3 notifies the Employer that the Contractor has defaulted.

Main Option clauses:

Option A:

36.3 accepts a quotation from the Contractor for acceleration and changes the Prices, Completion Date and the Key Dates
54.2 receives the Contractor's revision of the activity schedule
63.14 agrees with the Contractor to use rates and lump sums instead of Defined Cost.

Option B:

36.3 accepts a quotation from the Contractor for acceleration and changes the Prices, Completion Date and Key Dates
60.6 corrects mistakes in the bill of quantities
63.13 agrees with the Contractor to use rates and lump sums instead of Defined Cost.

Option C:

11.2(25) decides Disallowed Cost
20.3 receives advice from the Contractor on practical implications of the design and subcontracts
20.4 prepares with the Contractor forecasts of total Defined Cost for the Works
26.4 accepts or does not accept the Contractor's proposed Subcontract Data
36.3 accepts a quotation for acceleration and changes the Prices, Completion Date and the Key Dates
40.7 assesses the cost of repeating a test after a Defect is found
52.3 is allowed to inspect the Contractor's accounts and records
53.1 assesses the Contractor's share

53.3 makes a preliminary assessment of the Contractor's share at Completion

53.4 makes a final assessment of the Contractor's share

53.5 accepts a proposal by the Contractor to change the Works Information

54.2 accepts or does not accept a change to the Contractor's planned method of working

63.15 agrees with the Contractor to use the Shorter Schedule of Cost Components

97.4 assesses the Contractor's share after certifying termination.

Option D:

11.2(25) decides Disallowed Cost

20.3 receives advice from the Contractor on practical implications of the design and subcontracts

20.4 prepares with the Contractor forecasts of total Defined Cost for the Works

26.4 accepts or does not accept the Contractor's proposed Subcontract Data

36.3 accepts a quotation for acceleration and changes the Prices, Completion Date and the Key Dates

40.7 assesses the cost of repeating a test after a Defect is found

52.3 is allowed to inspect the Contractor's accounts and records

53.7 makes a preliminary assessment of the Contractor's share at Completion

53.8 makes a final assessment of the Contractor's share

53.5 accepts a proposal by the Contractor to change the Works Information

60.6 corrects mistakes in the bill of quantities

63.9 agrees with the Contractor to use bills of quantities in assessment of compensation events

63.13 agrees with the Contractor to use rates and lump sums instead of Defined Cost

63.15 agrees with the Contractor to use the Shorter Schedule of Cost Components

93.5 assesses the Contractor's share after certifying termination.

Option E:

11.2(25) decides Disallowed Cost

20.3 receives advice from the Contractor on practical implications of the design and subcontracts

20.4 prepares with the Contractor forecasts of total Defined Cost for the Works

26.4 accepts or does not accept the Contractor's proposed Subcontract Data

36.4 accepts a quotation for acceleration and changes the Prices, Completion Date and the Key Dates

36.5 accepts or does not accept a Subcontractor's proposal to accelerate

40.7 assesses the cost of repeating a test after a Defect is found

52.3 is allowed to inspect the Contractor's accounts and records

63.15 agrees with the Contractor to use the Shorter Schedule of Cost Components.

Option F:

11.2(26) decides Disallowed Cost

20.3 receives advice from the Contractor on practical implications of the design and subcontracts

20.4 prepares with the Contractor forecasts of total Defined Cost for the Works

26.4 accepts or does not accept the Contractor's proposed Subcontract Data

36.4 accepts a quotation for acceleration and changes the Prices, Completion Date and the Key Dates

52.3 is allowed to inspect the Contractor's accounts and records.

Secondary Option clauses:

W1 may agree with the Contractor to extend the time for notifying and referring a dispute

X2 notifies the Contractor of a compensation event for change in the law

X7 assesses the benefit to the Employer of taking over part of the Works

X13 accepts or does not accept a Contractor's performance bond

X14 accepts or does not accept the issuer of an advanced payment bond

X20 receives the Contractor's report of performance against Key Performance Indicators.

Supervisor actions

The Supervisor:

10.1 acts as stated in the contract and in a spirit of mutual trust and co-operation

13.1 replies to a communication within the period for reply

13.6 issues certificates to the Project Manager and the Contractor

14.2 may delegate any of his actions

40.3 notifies the Contractor of tests and inspections

40.5 does his tests without causing unnecessary delay

41.1 notifies the Contractor that Plant and Materials have passed tests

42.1 may instruct the Contractor to search for a Defect

42.2 notifies the Contractor of each Defect which he finds

43.3 issues the Defects Certificate

71.1 marks Equipment, Plant and Materials outside the Working Areas.

Notes

1 M. Latham (1994), *Constructing the Team: Final Report of the Government Industry Review of Procurement and Contractual Arrangements in the UK Construction Industry*. HMSO, London.

2 'NEC3 Engineering and Construction Contract' (2005). Thomas Telford Ltd, London.

3 'NEC3 Engineering and Construction Contract' (2010). Thomas Telford Ltd, London.

4 RICS (2007), *Contracts in Use: A Survey of Building Contracts in Use during 2007*. Royal Institution of Chartered Surveyors, London.

1 Early warnings

1.1 Introduction

Early warnings are covered within the ECC by Clauses 16.1 to 16.4.

Early warnings are an essential and valuable risk management tool within the NEC3 contracts, the ECC obliging the Project Manager and the Contractor to notify each other as soon as either becomes aware of any matter which could affect the project in terms of time, cost or quality.

Under Clause 16.1 the Contractor and the Project Manager give an early warning by notifying the other as soon as either becomes aware of any matter which could:

- *Increase the Total of the Prices.*
  The Price of the Works, in the form of the activity schedule, the bill of quantities or the target.
- *Delay Completion.*
  The Completion of the whole of the Works.
- *Delay meeting a Key Date.*
  The Completion of an intermediate 'milestone date' in accordance with Clause 11.2(9).
- *Impair the performance of the Works in use.*
  This sometimes causes confusion, but if we take as an example on an Employer-designed contract, the Contractor is instructed by the Project Manager to install a particular type of pump and the Contractor knows from experience that that pump would probably not be sufficient to meet the Employer's requirements once the works are taken over, then the Contractor should give early warning.

The Contractor may also give an early warning by notifying the Project Manager of any other matter which could increase his total cost. One could query whether a matter which could increase the Contractor's cost, but not affect the Price, should be an early warning matter, or for that matter, whether it should be anything to do with the Project Manager, particularly if Option A or B has been selected so the Contractor is not looking to recover this additional cost through the contract, but the words are 'the Contractor may give an early warning' so he is not obliged to do so.

Note that, also within Clause 16.1, the Contractor is not required to give an early warning for which a compensation event has previously been notified, so, as an example, if the Project Manager gives an instruction which changes the Works Information, it is a compensation event, for which neither the Project Manager nor the Contractor are required to give early warning.

The early warning procedure obliges people to be 'proactive', dealing with risks as soon as the parties become aware of them, rather than 'reactive' waiting to see what effect they have then trying to deal with them when it is often too late. Encouraging the early identification of problems by both parties puts the emphasis on joint solution finding rather than blame assignment and contractual entitlement.

It is one of the most important and valuable aspects of the contract and it is perhaps surprising that, whilst some other contracts refer to an early warning process, only the NEC contracts set out in clear detail what the parties are obliged to do, with appropriate sanctions should the parties not comply (see Clauses 11.2(25), 61.5 and 63.5).

1.2 Notifying early warnings

The contract requires (Clause 13.1) that all instructions, notifications, submissions, etc are in a form that can be read, copied and recorded, so early warnings should not be a verbal communication such as a telephone conversation. If the first notification is a telephone conversation, or a comment in a Site meeting, it should be immediately confirmed in writing to give it contractual significance.

Also, Clause 13.7 requires that notifications which the contract requires must be communicated separately from other communications, therefore early warnings must not be included within a long letter which covers a numbers of issues, or embodied within the minutes of a progress meeting.

There are some key words within the obligation to notify:

- 'The Contractor and the Project Manager' – No one else has the authority or obligation to give an early warning. The Project Manager is therefore notifying on behalf of himself, the Employer, the Supervisor, the Employer's Designers and many possible others whom he represents within the contract. The Contractor is notifying on behalf of himself, his Subcontractors, his Designers (if appropriate), and again many possible others whom he represents under the contract. Early warnings should be notified by the key people named in Contract Data Part 2.

  Project Managers are often criticised for seeing early warnings as something the Contractor has to do, and in fact most early warnings are actually issued by the Contractor. However, the Contractor and the Project Manager are obliged to give early warnings each to the other, so it is critical that Project Managers play their part in the process.

  As an example, if the Project Manager becomes aware that he will be late in delivering some design information to the Contractor, he should issue

Contract:	**EARLY WARNING NOTICE**
Contract No:	EWN No..

Section A: Enquiry

To : Project Manager/Contractor

Description

This matter could:

- ☐ **Increase the Total of the Prices**
- ☐ **Delay Completion**
- ☐ **Delay meeting a Key Date**
- ☐ **Impair the performance of the Works in use**

Risk reduction meeting called?   **Yes/No**   **Date: ....**

Signed:................................(Contractor/Project Manager)   **Date: ....**

Action by: **Date required:**

Section B: Reply

To: Contractor/Project Manager

Signed:................................(Contractor/Project Manager)   **Date: ....**

Copied to:

Contractor ☐ Project Manager ☐ Supervisor ☐ File ☐ Other .... ☐

Figure 1.1 Sample early warning notice

the early warning as soon as he becomes aware that the information will not be delivered to the Contractor, not wait and subsequently blame the Contractor for not giving an early warning stating that he has not received the information!

- 'As soon as' – means immediately. There are a number of clauses within the contract that deal with the situation where the Contractor did not give an early warning. Whilst the party who gives the early warning must do so as soon as he becomes aware of the potential risk, the other party should respond as soon as possible and in all cases within the period for reply in Contract Data Part 1.
- 'Could' – not 'must', 'will' or 'shall'. Clearly there is an obligation to notify even if it is only felt something may affect the contract, but there is no clear evidence that it will.

The Project Manager enters early warning matters in the Risk Register. Early warning of a matter for which a compensation event has previously been notified is not required. If the Project Manager gives an instruction for which a compensation event has already been notified, there is no requirement for either party to give an early warning.

It must be emphasised that early warnings are not the first step towards a compensation event as is often believed. Early warnings feature in a completely separate section of the contract and in fact the early warning provision is intended to prevent a compensation event occurring or at least to lessen its effect. It can also be used to notify a problem which is totally the risk of the notifier. It is also worth mentioning that early warnings are a notice of a future risk, not a past one. The parties are not required, nor is it of any value, to notify a risk that has already happened.

Example

Two weeks before the Completion Date the Contractor is informed by his Subcontractor that they have insufficient ceiling tiles to complete the Works and that this could potentially cause a delay in Completion.

Should the Contractor issue an early warning to the Project Manager bearing in mind that there has been a failure on the part of the Subcontractor, and probably the Contractor?

Yes, he should issue an early warning as it could delay Completion. Of course, there is no entitlement on the part of the Subcontractor or the Contractor in terms of a compensation event, but he should still issue the early warning.

The Project Manager and Contractor may possibly hold a risk reduction meeting, also inviting the Subcontractor to attend and they could consider

proposals and seek solutions, and decide action such as the Employer taking over the Works and the Contractor and Subcontractor returning to
Site when the material has been delivered, or the Employer taking over the
parts of the Works which are not related to this particular material. With a
degree of innovation and collaboration brought about by the early warning
provision the effect of this oversight could be mitigated.

Remedy for failure to give early warning

Under Clauses 61.5 and 63.5, if the Project Manager decides that the Contractor did not give an early warning of the event which an experienced Contractor
could have given, he notifies this to the Contractor when he instructs him to
submit quotations. If the Project Manager has done so, the event is assessed as if
the Contractor had given early warning.

Example

On 1 February, the Contractor becomes aware of an issue which could
increase the Total of the Prices and delay Completion. He should immediately give an early warning to the Project Manager; however he fails to do
so until 22 February, three weeks later.

When the Contractor gives the early warning, the Project Manager notifies a compensation event and instructs the Contractor to submit a quotation, at the same time notifying the Contractor that he did not give an early
warning of the matter that an experienced Contractor could have given.
If the Project Manager has given such notification, then the compensation event is assessed as if the Contractor had given early warning, so the
Contractor should price his quotation based on the event as at 1 February,
rather than 22 February.

Failure to give an early warning can also be considered as Disallowed Cost under
Options C, D and E, Disallowed Cost being the cost which the Project Manager
decides was incurred only because the Contractor did not give an early warning
which the contract required him to give.

1.3 Dealing with risk

Risks can be classified in an ECC contract as:

- risks which are owned by the Employer and do not affect the Contractor;
- risks which are owned by the Employer and do affect the Contractor

in terms of Price, Key Dates and/or Completion Date, and are therefore included and managed as compensation events;

- risks which are owned by the Contractor and are therefore deemed to have been included within his Price and programme; these are risks which may affect the Contractor but are not listed as compensation events.

Risks held by the Employer are listed under Clause 80.1, and if they occur they are compensation events under Clause 60.1(14). Clause 80.1 also allows for additional risks held by the Employer to be identified in the Contract Data.

All other risks are held by the Contractor.

Dependent on the Main Option chosen, a risk held by the Contractor may still be paid by the Employer, even though it will not be a compensation event.

General principles of risk management

There are a number of definitions of risk, an example being:

> Any uncertain event or set of events that, should it or they occur, would have an effect on achieving a party's objectives, in terms of presenting a negative threat or preventing a positive opportunity.

The objectives in a construction project are to build safely and to the required quality, to complete within the required time scale, and within the required budget.

Risk is always inherent in any activity, no matter how simple or complex it may be. However, the degree of risk will vary depending upon the circumstances involved and the amount of risk that one is willing to assume is normally directly proportional to the amount of reward that is anticipated. The greater the anticipated reward, the greater the risk one is normally willing to assume.

Risk is an integral part of the construction process and an important factor in the determination of construction costs and durations.

The parties to a construction contract assume that the contract will not only provide them with adequate protection, but also provide a fair and workable framework within which a vastly complex project can be constructed to the mutual satisfaction of all concerned.

The construction contract itself should be clear, concise, complete, legally correct and equitable. It should address the risks directly and attempt to eliminate as much of the ambiguity as possible.

Characteristics causing construction projects to present risks

All construction projects can be regarded as having risk as an underlying factor, though there are a number of differences between a construction contract and a manufacturing process, in that every construction project:

- is different, the only similarities being the people who manage them and the Materials and Equipment used;

- is not produced in a factory environment, but on a Site which is subject to fairly predictable seasonal variations, but unpredictable weather exceptions;
- brings a temporary multi-skilled and experienced team together, brought together for the purpose of constructing the project and possibly never working together again.

A thought-provoking quotation which, while written some time ago and in a 'comical' style, still holds a certain degree of truth:

> As teams go [the so-called project team] really is rather peculiar, not at all like a cricket eleven, more like a scratch bunch consisting of one batsman, one goal keeper, a pole vaulter and a polo player. Normally brought together for a single enterprise, each member has different objectives, training and techniques and different rules. The relationship is unstable, even unreliable, with very little functional cohesion and no loyalty to a common end beyond that of each member coming through unscathed.[1]

As a result of the above, one can easily draw the following conclusions:

- It is impossible to produce a design which will not require subsequent modifications, either because of Employer changes or because the design is found in practice not to work.
- It is impossible for a Contractor to produce a tender which will correspond exactly to the cost plus overheads and profit which will actually be incurred during the construction.
- It is impossible for a Contractor to produce a programme which can in all respects apply without revision and updating.
- When people are faced with such uncertainty they may, in an effort to defend their own beliefs and organisations, engage in conflict situations.

As a consequence, it is clear to see that both Employer and Contractor organisations are faced with risks that occur as a result of the nature of construction.

From the moment the decision to begin to redesign is taken until the new facility is in use, the Employer and Contractor take on board risks.

In addition to inherent risks within the construction risks there are also risks that will arise from the agreement between the parties. These contractual risks are primarily expressed within the contract as intended between the parties, but they can also arise through inexperienced drafting which can lead to lack of contract clarity, ambiguities and inconsistencies.

The main principles of apportioning risk in construction contracts are as follows:

- The Employer will always pay for risk, either by retaining the risk within the contract and paying for it should it occur, or the Contractor prices it within his tender.
- Risks should be apportioned to the party best able to manage and/or control them and to sustain the consequences if they materialise.

- Risks outside the Contractor's control and/or foreseeability should remain with the Employer.
- The party carrying the risk should be motivated to manage the risk, possibly through a risk/reward process.

The major types of risks that can arise on construction projects are as follows.

Risks to Employer

Technical:

- incomplete or incorrect design;
- inadequate Site investigations.

Logistical:

- availability of resources;
- availability of transportation.

Resources:

- insufficient resources and expertise;
- weather and seasonal problems;
- industrial relations problems.

Timing:

- inability to give access or possession.

Financial/contractual:

- inflation;
- delay in payment;
- non-availability of funds;
- cash flow problems;
- ambiguities and/or inconsistencies in contract documents;
- work carried out by directly employed parties;
- losses due to default of Contractors.

Safety/reliability:

- delays due to having to satisfy requirements.

Quality assurance/standards:

- quality issues.

Choice of Site/physical:

- ground conditions;
- ground water conditions;
- exceptional weather conditions;
- damage to neighbouring buildings;
- soil strata type and variability;
- contamination.

Environmental and planning:

- existing rights of way;
- public enquiries;
- planning requirements;
- noise abatement;
- availability of services.

Political:

- changes in Government/new legislation;
- customs and import restrictions.

Design:

- suitability and adaptability of the design.

Unfamiliarity:

- pioneer design;
- experience of design team.

Inflation:

- differential inflation due to market situation at time of tendering;
- abnormal inflation.

Errors in tender documents

Construction:

- bankruptcy/insolvency;
- industrial action.

Variations:

- effect on construction duration and Price of late variations;
- effect of number of orders.

Construction delays:

- delay by specialist Contractors and statutory authorities;
- delay in giving instructions to the Contractor.

Force majeure:

- fundamental risk, which cannot be priced.

The risk management process

Risk management is a structured approach to identifying, assessing and managing risk.

It can be defined as:

> The systematic application of management policies, procedures and practices to the tasks of identifying, analysing, assessing, treating and monitoring risks.

The risk management process can help a party analyse the degree of uncertainty in achieving the contract/project objectives, understand the consequences of particular events occurring, and develop management actions to control the event or minimise its consequences.

There are six primary activities in the process.

(i) Risk identification

Risks should be identified as early as possible and continuously monitored to highlight new risks and the passing of old risks. Risk identification should initially be carried out by analysing all available documentation and knowledge as a brainstorming session involving a number of individuals, ideally chaired by an independent person external to the project. This will ensure maximum and unbiased contributions from all parties.

(ii) Risk analysis

Once the risks have been identified they should be classified in terms of:

- the likelihood of occurring;
- the impact on the project objectives should they occur.

Each risk can then be given a ranking, possibly:

- extremely unlikely;
- unlikely;
- moderate;
- likely;
- almost certain.

Note that 'will not occur' and 'certain to occur' are not classifications or we would then be considering factual certainties rather than possible risks.

Impact can also be given a ranking, possibly:

- insignificant
- moderate
- major
- severe
- catastrophic.

(iii) Risk planning

This is the process of developing responses to address, manage or control risks.
   The responses will consider:

- *avoiding the risk* – the Employer abandoning the project, the Contractor deciding not to tender;
- *reducing the probability* – doing something a different way or educating the people who may be subject to the risk, so the risk is less likely to occur;
- *reducing the impact* – including resources, funds and/or float in programmes to cushion the effect of the risk;
- *transferring the risk* – passing the risk to another (willing) party or an insurance company who is best able to manage or bear the risk;
- *accepting the risk* – with the full knowledge of what the risks and potential outcomes are;
- *ignoring the risk* – with the full knowledge of what the risks and potential outcomes are.

(iv) Risk tracking

Risks should be tracked throughout the project to monitor the status of each risk as the project progresses. This is normally done through a Risk Register.

(v) Risk controlling

An activity that utilises the status and tracking information about a risk or risk mitigation effort. A risk may be closed or watched; a contingency plan may be involved.

(vi) Risk communicating

An action to communicate and document the risk at all times. Again, this will normally be done using a Risk Register.
   When considering risk it is vital to consider:

- which party can best foresee the risk;
- which party can best bear the risk;
- which party can best control the risk;
- which party most benefits or suffers if the risk materialises.

The use of a Risk Register under the ECC

It is critical that the parties fully understand that the purpose of a Risk Register is to list all the identified risks and the results of their analysis and evaluation. It can then be used to track, review, monitor and communicate risks as they arise to enable the successful Completion of the project. The Risk Register does not allocate risk – that is done by the contract; rather, it records the risk.

Clause 11.2(14) defines the Risk Register as 'a register of the risks which are listed in the Contract Data and the risks which the *Project Manager* or the *Contractor* has notified as an early warning matter. It includes a description of the risk and a description of the actions which are to be taken to avoid or reduce the risk'.

In that sense, the contract does not prescribe the format or layout of the Risk Register, or its intended purpose, other than to list the risks in the contract and those that come to light at a later date and are notified as early warnings following which, if there is a risk reduction meeting, the Project Manager revises the Risk Register to record the outcome of the meeting.

Note that by the definition with Clause 11.2(14), risks which were not originally included in the Contract Data or subsequently notified as early warnings should not be included in the Risk Register. Contract Data Part 2 allows the Contractor to identify matters which will be included in the Risk Register. The Risk Register does not allocate or change the risks in the contract, it records them and assists the parties in managing them. In that sense it is a valuable addition to the contract which was absent from previous editions but introduced in NEC3.

Figure 1.2 shows an extract from a typical register which would comply with the contract in that it includes the basic requirement for a description of the risk and a description of the actions which are to be taken to avoid or reduce the risk.

There is no stated list of components of a Risk Register, but column headings should typically be titled:

- *Description of risk*
  A clear description of the nature of the risk, if necessary referring to other documents such as Site investigations, etc.
- *Implications*
  What would happen if the risk were to occur?
- *Likelihood of occurrence*
  This provides an assessment of how likely the risk is to occur. The example shows a forecast on a 1 (least likely) to 5 (most likely) basis, though it may be assessed as percentages, colour coding, or simply 'Low' (less than 30 per cent likelihood), 'Medium' (31–70 per cent likelihood), 'High' (more than 70 per cent likelihood).
- *Potential impact*
  This assesses the impact that the occurrence of this risk would have on the project in terms of time and/or cost. The example shows the assessment on a 1 (low) to 5 (high) basis.

RISK REGISTER

Contract:...............

Contract No:...............

Contractor...............

Project Manager...........

Description of Risk	Implications	Likelihood of Occurrence (1–5) (Least – Most)	Potential Impact (1–5) (Low – High)	Risk Score	Risk Owner	Mitigation Strategy By whom? By when?	Allowance in the Total of the Prices	Programme Allowance	Cost Allowance	Risk Status	Last Updated
Discovery of unforeseen existing foundation bases	Delay and additional costs in removal	2	4	8	Initially Contractor unless additional silo bases discovered then Employer as comp. event	Site Information identifies likely locations Contractor to take due regard	No allowance for unforeseen	No allowance for unforeseen	£20,000	Reducing Excavation 50% complete No additional silo bases discovered to date	15/6/2011

Figure 1.2 Extract from a Risk Register

- *Risk score*

  Risk = Likelihood of something occurring × Impact should it occur.

  By evaluating risk on that basis each can be evaluated and categorised into an order of importance. This formula can only be applied to economic loss, and different standards would need to be adopted when considering risk of death or serious bodily injury.
- *Risk owner*

  This identifies which party or individual owns the risk, this is allocated through the contract, the principle being that if the Contractor owns it, he is deemed to have allowed for its Price and/or time effect within his tender. If the Employer owns it, it will be a compensation event.
- *Mitigation strategy*

  This registers what actions are proposed which could be taken to prevent, reduce or transfer the risk. Also, who is responsible for this action, and when should the action take place?
- *Allowance in the Total of the Prices*

  How much has the Contractor allowed in his tender for a risk he owns?
- *Programme allowance*

  How much time has the Contractor allowed in his programme for a risk he owns?
- *Employer cost allowance*

  If it is an Employer's risk, how much has he allocated to cover a potential compensation event?
- *Risk status*

  This identifies whether the risk is current, and also whether it is increasing, decreasing or has not changed since it was last reviewed?
- *Last updated*

  When was an additional risk identified? When was the Risk Register last updated?

In effect, the Contractor could seek to qualify his tender by including additional matters within Contract Data Part 2 which he sees, or would like to transfer as Employer's risks.

Risk management is essential to the success of any project and normally follows five steps:

1. Identify the risks.
2. Decide who could be harmed and how.
3. Evaluate the risks and decide on precautions.
4. Record the findings and implement them.
5. Review the assessment and update if necessary.

In order to compile the Risk Register the risks must first be listed, then they are quantified in terms of their likelihood of occurrence and their potential impact upon the project.

Risk reduction meetings within the ECC

Under Clause 16.2, either the Project Manager or the Contractor may instruct the other to attend a risk reduction meeting.

The key word in Clause 16.2 is 'instruct'. If the Contractor calls a risk reduction meeting, he is instructing not merely requesting the Project Manager to attend. This clause is clearly intended to promote ownership of the project by the Contractor and the Project Manager, and any issues that could affect it, together with their resolution.

A risk reduction meeting need only have the Project Manager and the Contractor present, though each may instruct other people to attend if the other agrees, so in reality a number of people normally attend the meeting. The 'rule of meetings' will often apply in that the productivity of the meeting is often inversely proportional to the number of people who attend it!

Clause 16.3 describes what the attendees at a risk reduction meeting should do in an effort to either resolve the problem or at least attempt to resolve it:

- Make and consider proposals for how the effect of the registered risks can be avoided or reduced ('registered' refers to the Risk Register).
- Seek solutions that will bring advantage to all those who will be affected.
- Decide on the actions which will be taken and who, in accordance with this contract, will take them.
- Decide which risks have now been avoided or have passed and can be removed from the Risk Register.

Clearly, the purpose of the risk reduction meeting is to actively consider ways to avoid or reduce the effect of the matter which has been notified. In some cases, the matter can be fully resolved, but in others, as the matter may not yet have occurred, it may simply have to be 'parked' and recorded as such in the Risk Register. It is the Project Manager's responsibility to revise the Risk Register to record the outcome of the meeting, and to issue the revised Risk Register to the Contractor.

Note

1 Architect Denys Hinton, speaking at the RIBA, London, in 1976.

2 Design

2.1 Introduction

Design, and more specifically Contractor's design, is covered within the ECC by Clauses 21.1 to 23.1, with further references within the Main Option clauses.

2.2 Some principles of design and design management

Good design and the management of the design process is an often underrated, but critical component in the success or failure of construction projects.

It is imperative that the Employer initially understands what he wants and is able to clearly communicate that to the Designer and other parties in the form of a clearly structured brief. The Designer can then convert that brief into a practical and economic solution which provides the deliverables of meeting quality standards, within budget and available time constraints.

Whilst quality and financial constraints tend to be uppermost in Employers' minds, it is critical that sufficient time is allowed for the design to be developed and alternatives considered, then the often time-consuming process of submission and awaiting regulatory approval, before making final decisions. In the European Union, consideration must also be given when designing projects in the public domain to time scales to comply with *Official Journal of the European Union (OJEU)* regulations.

Key points in effective design management are as follows:

- Employers should always be clear as to their objectives, priorities and operational requirements, including the needs of all stakeholders, before appointing Designers, and ensure that these are clearly stated in their brief.

  The initial stages are the most important part of any project, and a well-informed Employer is the critical foundation to the successful delivery of the project within the budget, agreed time scale and to the required standards.

- Allocate sufficient time and resources to establish the Employer's design quality aspirations ensuring that design time is not reduced in order to recover time lost in the programme for other reasons. Good design always takes time. Time allocation should also include the need for external regulatory approvals over which the Employer and his Designers may have little or no control.

- Appoint someone responsible for taking the lead on design issues and co-ordinating the work of the various parties, again including regulatory bodies.
- Continue to communicate design requirements throughout the design, procurement and construction stages.
- Good design should always consider the most economical solution, both in terms of initial cost, but also life cycle cost, including future maintenance and removal and replacement; therefore value engineering and continuous design reviews are integral to the process.
- Good design should consider not only aesthetic and structural features, but also functionality and efficiency, build quality, accessibility, sustainability, whole life costing and flexibility in use.
- Employers should always be clear as to what funding they have available for the design process and for what is to be designed. That funding should always be realistic.
- A candid and honest, but formal post-Completion audit should always be carried out to ascertain the successes and failures of the project from which lessons can be learned and carried through to the next.

2.3 Selection of the design team

There are several methods of selecting the appropriate design team for a particular project including, particularly for a high-profile project, design competitions, which primarily select specific design ideas or outline designs for a project, rather than the individual design team members.

The key issue with design competitions is that the Employer must have a clear vision of what he requires in order that design teams will have clear criteria from which to create and submit their designs.

Design competitions may be appropriate where there is either a unique problem that will benefit from a wide range of design approaches being explored (along with likely considerable public interest or resistance – which may be the case on a major new public building) or where the competition promoter wishes to encourage the development of new talent.

Alternatively, one can focus on the individual Designers and their skills and reputation in designing similar projects. Again, it is vital that not only aesthetic qualities are considered, but also the design team's ability to perform within the time constraints and budget, ie to manage the design.

With the European Union, all public sector appointments, irrespective of the Employer's preferred nature of competition or reference to any other guidance on design competitions, must be consistent with EC procurement rules in terms of process and outcome.

Whichever process is used for selecting the design team, there must be clear and objective criteria on which to select and appoint the successful Designers and this should consider:

- the balance between quality, cost and time-related issues;
- the technical ability of the design team to deliver the project on time and within budget, as well as to the required standards of quality;
- the financial standing of the Designers including insurance requirements, etc. The financial failure of a Designer can be catastrophic to the successful Completion of the project.

2.4 Contractor's design

The NEC3 is unique amongst construction contract families in that, whilst most other contracts provide for portions or all of the design to be carried out by the Contractor, they all use a separate contract for design and build where the Contractor has full responsibility for design, whereas the ECC uses the same contract whether or not design is to be carried out by the Contractor as any design criteria and obligations are defined within the Works Information.

In reality, there is no need for a separate design and build contract as within the ECC the clauses which cover early warnings, programme, payment, change management, etc are the same whether or not the Contractor has design responsibility.

The 'default position' within the ECC is that all design is carried out by the Employer, but any design to be carried out by the Contractor is assigned to him within the Works Information and Clauses 21 to 23 cover design obligations and submission and acceptance of the Contractor's design proposals.

The most important advantage with design and build is that the Contractor is wholly responsible for the design. This 'single-point' responsibility is very attractive to Employers, particularly those who do not have the knowledge or experience to appoint design consultants, or who may not be interested or able to distinguish the difference between a design fault and a workmanship fault. However, it is vital that the Employer is able to detail his requirements clearly.

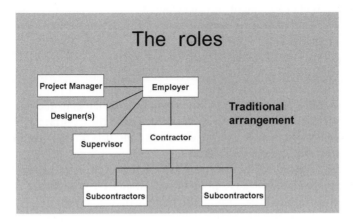

Figure 2.1a The roles: traditional arrangement

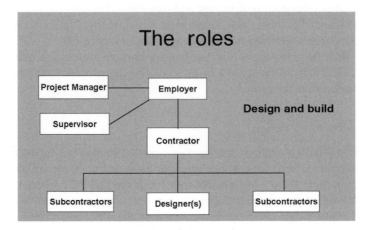

Figure 2.1b The roles: design and build

This single-point responsibility also means that the Contractor is not relying on other firms (eg Architects) for the execution of design or for the supply of information.

The design and build process increases the opportunities for harnessing the benefit of the Contractor's experience during the design stages of the project. Many of the developments in procurement processes, and much of the work in the field of study known as 'buildability', have had this purpose in mind. The benefits of the integration of Designers and builders are more economically constructed projects, as well as a more economic and effective production process.

The technical complexity of the Employer's process may lead to solutions which are more akin to engineering than to building. While this does not mean that design and build is unsuitable, it is likely that in such circumstances the use of other forms of procurement will be favoured.

The reasons for allocating some or most of the design to the Contractor can then be summarised as:

Advantages

- Single-point responsibility, so the traditional design/construction interface and the risks associated with it are transferred to the Contractor. Whether a Defect is a design fault or a workmanship fault, it is the responsibility of the Contractor to correct that Defect.
- It is not necessary for the design and build Employer to be an 'expert' Employer and appoint and manage a group of consultants. He engages the Contractor as a single point of responsibility to provide that expertise.
- The management of the design risk by the Contractor can result in greater certainty of the time, cost and project objectives being met.

- The procurement period is shorter as there is no requirement for the Employer to complete the design and prepare bills of quantities etc prior to appointing the Contractor.
- The project can be completed on Site quicker as design and construction can overlap.
- The Contractor can utilise his experience and expertise in using preferred methods of construction to minimise time and cost.
- As the Contractor has responsibility for design and construction, the design should be more rational and buildable.

Disadvantages

- The Contractor designs and builds to a Price commitment; therefore where there is a conflict between aesthetic quality and ease of fabrication, the requirements for fabrication will override.
- The Employer has reduced control of design and quality as these are passed to the Contractor at an early date.
- The Employer has reduced control of change costs as there is no detailed bill of quantities to relate to.
- It may not be good for complex projects because of the more comprehensive pre-contractual documentation which may be required by the Employer.
- Whilst the procurement period may be shortened, the Contractor must be given more time to prepare his tender.
- It is expensive for the Contractor to prepare his tender as he needs to consult with Designers as well as to price the tender.

If the Contractor is to design all or part of the Works, the Employer must create a clear and unambiguous set of requirements, clearly defining what he requires the tendering Contractors to price and eventually to carry out. The Contractor responds with his proposals which will include production as well as design work.

The Contractor's design input varies from one contract to another, ranging from the mere development of a detailed design which was provided by the Employer, to a full design process including proposals, sketch schemes and production information.

There will often be some negotiation between the Employer and the Contractor, with the aim being to settle on an agreed set of Contractor's proposals. These proposals will include the contract Price, as well as the manner in which it has been calculated.

Once the Employer's requirements and the Contractor's proposals match, the contract can be executed and the Contractor can implement the work.

One reported disadvantage of design and build is where there is a conflict between aesthetic quality and ease of construction, as the Contractor is also designing the project, the requirements for construction will override.

A further criticism has been that a design and build Contractor will put in the minimum effort with regard to the design in an effort to secure the contract. These two criticisms suggest strongly that quality, particularly design quality, will suffer under this procurement process. However, this is not a valid criticism of design and build as a procurement form, but rather of the people who advise Employers and prepare their requirements and tender documents.

When selecting a Contractor, the Employer may approach design and build Contractors who will prepare the design and submit their tenders based on a 'shopping list of requirements', or in many cases will employ his own Designers to carry out initial designs of major elements of the Works, which would then form the basis of the Employer's requirements, with the Designers transferred to the Contractor by a novation agreement.

2.5 Incorporating Contractor's design into the ECC

Establishing the Employer's Works Information

In the ECC 'the Contractor Provides the Works in accordance with the Works Information' (Clause 20.1) and 'the Contractor designs the parts of the Works which the Works Information states he is to design' (Clause 21.1).

It is clear, then, that the Works Information should clearly show what is to be designed by the Contractor and this could consist of detailed specifications or some form of performance criteria and certain warranties.

When preparing the Works Information, which includes Contractor's design, it is important that the right balance of information should be considered. If it is too prescriptive it will lead to all the tendering Contractors submitting similar designs and Prices. If the Works Information is not detailed enough, then none of the tendering Contractors will produce a design which is as the Employer intended.

As an example, the Employer wishes to invite tenders for the design and build of 52 houses. The Employer has certain essential requirements, but wishes to attract some innovation in that Contractors will interpret the shape, size and appearance of the houses in different ways subject to the approval of external bodies such as the local authority planners and other legislative bodies, etc.

The Works Information should therefore comprise the following groups:

- *Essentials* – Requirements that the Employer must have. For example, if the Contractor is to design and build houses then the number of houses and the mix of types of houses, for example number of bedrooms, should be clearly stated. Any other essentials should also be clearly stated, otherwise a tenderer may not include them.
- *Preferences* – Requirements that the Employer would prefer to have. For example, the Employer may have a preference for a certain heating system, but the Contractor may offer an alternative which is more efficient, cheaper to run, etc.

The Employer may also state within his preferences that 'equal and approved' may be acceptable subject to the Contractor proving that it truly is 'equal' and also that he is prepared to approve it. As part of this proof the Contractor should consider issues such as life cycle costing. For example, if the Contractor can prove that the alternative heating system is equally efficient and easy to operate, but in the longer term the alternative requires a higher level of maintenance then the Employer will need to be made aware of this requirement. It is suggested that the preferences should not apply to critical aspects of the Project.

- *Innovations* – Something the Employer may not have thought of that the Contractor believes could win the tender for him!

This is open to the Contractor to provide as added value to his particular bid. For instance, one Contractor may offer a children's playground, or particularly 'green' design/construction over and above the requirements of the Works Information and/or the relevant legislation. If a Contractor feels he can exceed the Works Information he should submit his proposals which comply with the document (a compliant bid) but list separately how he could exceed it and any time and/or Price implications.

With regard to the Contractor's proposals, he will wish to demonstrate that his bid is compliant with the Works Information, though it is important that he is not prescriptive about exactly what he will provide, as his design may evolve and develop as the construction progresses. He may find new Equipment and Materials, or he may wish to use some monies to offset any problems that may arise.

As an example, the Employer wishes to procure a new office block on a full design and build basis.

Typically he should consider the following:

Quality

What are his overall requirements?

- The size and use of the building. If no specific size of building is known, details of exactly what the Employer intends to use the proposed building for and expected number of occupants.
- Open plan or cellular offices.
- Any specific requirements, for example canteen facilities, gym, etc. Number of floors, clear span, or sub-divisions.
- Need for future flexibility and expansion.
- Specific heating requirements, lighting levels for various parts of the building, air conditioning requirements. Running costs and energy conservation requirements.
- Car parking facilities.

- Maintenance, running costs and life cycle cost factors. How long is the building expected to last?

Time

When will the Employer need the building?

- Is the starting date and/or Completion Date critical to the Employer?
- Consider pre-tender activities such as planning consents, preparation of tender documents and also other operations which need to be carried out before the Contractor can be appointed and start on Site.

Cost

How much will be paid?

- What is the budget for the building? How restricted is that budget? Is there a cost plan and expected cash flow projections for the building? How long will it take to prepare one?
- Is the project being wholly funded by the Employer or is partial funding coming from another source, for example grant aid? How much influence do the funders have in terms of the design and construction of the building?

From these initial considerations the documentation to be prepared for tender can be established:

- What drawings are to be included within the Works Information? Broad sketch plans, showing size and general layout only? Or detailed design drawings?
- Site Information. How detailed will it be? Who is responsible for accuracy?
- Who will apply for outline and later detailed planning consents? Possible delays with consents? Revised applications due to changes in design suggested by Contractor?
- Performance specifications? Or specification of Works and Materials? Outline specifications or detailed at this stage?
- Any major products, specialist or supply items over which the Employer wishes to exercise some choice or control. Nomination is difficult with design and build and creates complications.
- Employer to provide an outline time programme, or left to the Contractor to submit with his tender?
- Sectional Completions? Split responsibilities and insurances.
- Constraints: access, space, working hours, sequence of working.
- Which Main and Secondary Options are to be used?
- Design and construction liability. Any special insurance requirements or performance bonds? Design warranties?

From the above it can be readily seen that it is vital that the Employer defines how much information and detail is required in the returned tender and it is crystallised within the Works Information. For example, a full set of drawings and comprehensive specification would not normally be expected from tenderers at tender stage and the scope and limitation of these documents must be defined.

The Contractor's Works Information

In simple terms, the Contractor's Works Information should be fully compliant with the Employer's Works Information. If the Employer's Works Information is detailed, the Contractor's Works Information should be detailed; if they are vague, the Contractor's Works Information may also be vague.

The Contractor's Works Information within his tender should incorporate the following:

- plans, elevations, sections or typical details;
- information about the structural design;
- outline drawings of incorporated services;
- specifications for Materials and workmanship;
- any alternative solutions departing from the brief;
- a method statement and construction programme;
- the Price;
- a staged payment schedule (if appropriate);
- proposed main Subcontractors and Suppliers.

Once the Employer's Works Information and the Contractor's Works Information are consistent, the contract can be executed and the Contractor can carry out the work.

It is important to note that if the Contractor's Works Information subsequently fall short of the Employer's Works Information then the Employer's Works Information will prevail.

Employer's Works Information v Contractor's Works Information

The ECC is unlike most other construction contracts in that it does not provide a priority or hierarchy of documents, but one can 'read into' the various clauses to determine which takes precedence in the event of an inconsistency.

A question to then be considered is, if the Contractor has full design responsibility and there is an inconsistency between the Works Information provided by the Employer and the Works Information provided by the Contractor, which takes precedence?

The answer can be found in Clauses 11.2(5) and 60.1(1).

Defects are defined in Clause 11.2(5) as:

- a part of the Works which is not in accordance with the Works Information; or
- a part of the Works designed by the Contractor which is not in accordance with the applicable law or the Contractor's design which the Project Manager has accepted.

Clause 60.1(1):

The Project Manager gives an instruction changing the Works Information except:

- a change made in order to accept a Defect; or
- a change to the Works Information provided by the Contractor for his design which is made either at his request or to comply with other Works Information provided by the Employer.

2.6 Acceptance of design

The Contractor is required to submit the particulars of his design as the Works Information requires to the Project Manager for acceptance. The particulars of the design in terms of drawings, specification and other details must clearly be sufficient for the Project Manager to make the decision as to whether the particulars comply with the Works Information and also, if relevant, the applicable law.

The Works Information may stipulate whether the design may be submitted in parts, and also how long the Project Manager requires to accept the design. Note that in the absence of a stated time scale for acceptance of design, the 'period for reply' will apply.

Many Project Managers are concerned that they have to 'approve' the Contractor's design and therefore they are concerned as to whether they would be qualified, experienced and also insured to be able to do so. Note the use of the word 'acceptance' as distinct from 'approval'. Acceptance denotes compliance with the Works Information or the applicable law; it does not denote that the design will work, that it will be approved by regulating authorities, or that it will fulfil all the obligations which the contract and the law impose, therefore performance requirements such as structural strength, insulation qualities, etc do not need to be considered.

A reason for the Project Manager not accepting the Contractor's design is that it does not comply with the Works Information or the applicable law.

The Contractor cannot proceed with the relevant work until the Project Manager has accepted the design.

Note that under Clause 14.1, the Project Manager's acceptance does not change the Contractor's responsibility to Provide the Works or his liability for his design. This is an important aspect as, if the Project Manager accepts the Contractor's design but following acceptance there are problems with the proposed design, for example in meeting the appropriate legislation or the requirements of external regulatory bodies, this is the Contractor's liability.

Under Clause 27.1, the Contractor obtains approval of his design from Others where necessary. This will include planning authorities, and other third party regulatory bodies.

Note also, Clause 60.1(1) second bullet point, which clearly refers to the fact that the Employer's Works Information prevails over the Contractor's Works Information.

This clause gives precedence to the Works Information in Part 1 of the Contract Data over the Works Information in Part 2 of the Contract Data. Thus the Contractor should ensure that the Works Information he prepares and submits with his tender as Part 2 of the Contract Data, complies with the requirements of the Works Information in Part 1 of the Contract Data.

2.7 Design of Equipment

If the Contractor has designed an item of Equipment, for example a temporary access way, or specialist scaffold, the Project Manager may instruct him to submit particulars to him for acceptance, so the submission is not obligatory as with the design of parts of the Works.

The Project Manager may not accept the design if it does not comply with the Works Information, the Contractor's design which the Project Manager has accepted, or the applicable law.

2.8 Intellectual Property Rights

The question of Intellectual Property Rights (IPR), which includes copyright, patents and trademarks, needs to be considered carefully where the Contractor is designing all or part of the Works.

The design and construction of a building or other structure is unique and unless the contract states otherwise, even though the owner pays for the design work, he does not automatically own the copyright to the design and the design documents. The Designer retains ownership and control over the design and the various documents, based on copyright law and the terms of agreement between the Employer and the Designer.

This copyright issue can be very complex as it only applies to that which was originally conceived by the Designer and therefore much of what is included in the design may not be subject to copyright, for example the layout of kitchen units or sanitary appliances within a bathroom.

Copyright also extends to copying the design, photocopying and distributing drawings and other documents, etc.

In the event that copyright is breached, the owner of the copyright may obtain an injunction preventing further breaches and, if necessary, may seek impounding of Materials, and also damages, loss of profits and legal fees. If a building design copyright has been breached, the Employer may be prevented from commencing or continuing with the Works.

2.9 Collateral warranties

A collateral warranty is an agreement associated with another contract, collateral meaning 'additional but subordinate' or 'running side by side', the collateral warranty being entered into by a party to the primary contract, for example a Contractor or a Subcontractor and a third party who is not a party to the primary contract but who has an interest in the construction project.

Collateral warranties are a relatively recent feature of contracts, coming into use in the past 20 years following some notable court cases in which it was judged that it was not possible to recover damages for negligence in relation to Defects in construction projects as it was held to be economic loss, which is not recoverable in tort.

Consequently, such claims had to be brought as claims for breach of contract, where economic loss is recoverable. It therefore became necessary to establish a contract between parties involved in a construction project, such as Designers or Subcontractors and parties such as funders, present and future purchasers, who are not parties to the primary contracts but who have an interest in the project, hence the 'collateral warranty'.

The NEC3 contracts do not provide for collateral warranties, though it is worth some commentary on their use and how they could be included specifically within the ECC.

Collateral warranties are different to traditional contracts, in that the two parties to the warranty do not have any direct commercial relationship with each other. They are useful for creating ties between third parties, but the main disadvantage of collateral warranties is the expense of providing them, particularly where there are many interested third parties such as with housing and commercial properties. When one also considers that parties can change their names or business status during the life of a warranty then even a medium-sized project can involve 30 or more warranties.

Different parties will also have different requirements for collateral warranties, therefore the precise terms of the collateral warranty agreement need to be considered in each case carefully so that they are drafted to provide the required protection.

Collateral warranties are executed as a 'deed' also referred to as 'under seal', which differs from a simple contract in two respects:

- First, in English law the time after a breach of contract has occurred within which one party can sue another is 12 years, whereas the time for a simple contract executed 'under hand' is 6 years.
- Second, whereas under English law consideration is needed for a contract to be effective, this is not the case with a deed. For a company to execute a deed effectively, the document must be signed by two directors or a director and the company secretary, unless the Articles of Association contain some other requirement.

The basic ingredients of a collateral warranty are usually:

(i) Definitions

There will be a brief description of the Works and the collateral warranty. The 'contract' is the contract to which the collateral warranty agreement is collateral and the date should be specified. Since the obligations of the Contractor to the Beneficiary parallel the obligations of the Contractor to the Developer under the contract for the Works, the Beneficiary should always ask to see a copy of the contract. If, for example, there are restrictions on the Contractor's liability in the principal contract, that might affect the rights of the Beneficiary under the collateral warranty.

A copy of the principal contract may be attached to the collateral warranty agreement.

(ii) The warranty

The Contractor gives warranties to the Beneficiary that he has exercised and will continue to exercise reasonable skill and care in his obligations to the Employer under the principal contract including compliance with any design obligations, performance requirements and correction of Defects.

The warranty should contain a statement that the Warrantor shall have no greater liability to the Beneficiary than they would have under the primary contract.

The Contractor also does not have any liability to the Beneficiary if he fails to complete the Works under the contract on time, as that will usually be covered by delay damages under the principal contract.

There should be a statement that the Warrantor shall have no liability under warranty in any proceedings commenced more than 12 years after the date of Completion of the Works.

(iii) Naming of the parties

The named parties can be, for example, the Contractor and the end user, a Sub-contractor and the Employer, or a Consultant and the Employer.

The party that gives the warranty is known as the Warrantor, the other party is known as the Beneficiary.

(iv) Prohibited Materials

The Contractor may be required to give a warranty that he will not specify or use Materials which are known to be prohibited or deleterious to health and safety.

In addition, the use of deleterious Materials may contravene relevant legislation or regulations as well as being expressly prohibited in the contract between the Developer and the Contractor.

(v) Copyright

Normally, the Beneficiary under a collateral warranty agreement will have a right to use designs and other documents prepared by the Contractor but only in connection with the project for which those design documents are prepared, copyright to these documents remaining with the Contractor.

(vi) Insurances

There is normally a requirement for the Warrantor to maintain insurances including Professional Indemnity (PI) insurance where applicable for a stated minimum level of cover and for a period of 12 years after Completion of the Works.

(vii) Assignment

There is normally a provision allowing the Beneficiary up to two assignments of the warranty.

(viii) Step-in rights

Step-in rights can allow a funder to take over the Employer's role and become the Employer under the primary contract. This is particularly important where the Employer is unable to pay the Contractor, the Contractor having to notify the Beneficiary before he terminates the contract, thus allowing the Beneficiary to 'step in' and take the Employer's place, with an obligation to pay the Contractor any money that is outstanding. The Beneficiary does not have any obligation to step in, only the right to do so. As the effect of this clause is to vary the principal contract it is important that, when the collateral warranty contains such a clause, the Employer should be a party to that document.

(ix) Law and jurisdiction

It is important to specify the law of the contract and the warranty and the jurisdiction of the courts that will have jurisdiction.

(x) Dealing with planning permission

The responsibility for obtaining planning permission can lie with the Employer or can be assigned to the Contractor.

In some cases, in a design and build contract, the Contractor may be responsible for gaining outline planning for the project; however, this should be a two-stage process so that if outline planning permission is not obtained, the second stage would not proceed. It is more usual in design and build for the Employer to gain outline planning permission and the Contractor to be responsible for gaining full and detailed planning consent.

2.10 Novation of Designers

Whilst the principle of design and build agreements is that the Employer preparers his requirements and sends them to the tendering Contractors and they prepare their proposals to match the Employer's requirements, in reality in nearly half of design and build contracts the Employer has already appointed a design team which prepares feasibility proposals and initial design proposals before tenders are invited. Outline planning permission and sometimes detailed permission may have also been obtained for the scheme before the Contractor is appointed.

Each Contractor then tenders on the basis that the Employer's design team will be novated or transferred to the successful tendering Contractor who will then be responsible for appointing the team and completing the design under a new agreement.

This process is often referred to as 'novation', which means 'replace' or 'substitute' and is a mechanism where one party transfers all its obligations and benefits under a contract to a third party. The third party effectively replaces the original party as a party to that contract, so the Contractor is in the same position as if he had been the Employer from the commencement of the original contract. Many prefer to use the term 'consultant switch' where the design consultant 'switches' to work for the Contractor under different terms as a more accurate definition.

This approach allows the Employer and his advisers time to develop their thoughts and requirements, consider planning consent issues, then when the design is fairly well advanced, the Designers can then be passed to the successful design and build Contractor.

The NEC3 contracts do not include pro forma novation agreements, but, if used, it is critical that the wording of these agreements be carefully considered as there are many badly drafted agreements in existence. Novation can be by a signed agreement or by deed. As with most contracts, there should be consideration, which is usually assumed to be the discharge of the original contract and the original parties' contractual obligations to each other. If the consideration is unclear, or where there is none, the novation agreement should be executed as a deed.

In many cases the agreement states briefly (and very badly) that from the date of the execution of the novation agreement, the Contractor will take the place of the Employer as if he had employed the Designers from the beginning, the document stating that in place of the word 'Employer' one should read 'Contractor'; however, the issues of design liability, inspection, guarantees and warranties may not always apply on a back-to-back basis, so one must take care to draft the agreement in sufficient detail and refer to the correct parties.

In practice, the best way is to have an agreement drafted between the Employer and the Designer covering the pre-novation period, and a totally separate agreement drafted between the Contractor and the Designer for the post-novation period.

When a contract is novated, the other (original) contracting party must be left in the same position as he was in prior to the novation being made. Essentially, a novation requires the agreement of all three parties.

As this book is intended as a guide for NEC3 practitioners worldwide, it is not intended to examine specific legal cases within the UK, as they may not apply on an international basis, but there is certainly case law available in terms of design errors in the pre-novation phase being carried through as the liability of the Contractor in the post-novation stage, and also the accuracy and validity of Site investigations and the Employer's and the Contractor's obligations in terms of checking and taking ownership of such information.

2.11 Defects in the Contractor's design

Defects are defined in Clause 11.2(5) as:

- a part of the Works which is not in accordance with the Works Information; or
- a part of the Works designed by the Contractor which is not in accordance with the applicable law or the Contractor's design which the Project Manager has accepted.

2.12 Use of the Contractor's design

Under Clause 22.1, the Employer may use and copy the Contractor's design for any purpose connected with construction, use, alteration or demolition of the Works unless otherwise stated in the Works Information and for other purposes as stated in the Works Information.

If the Employer wishes to own the rights to any design material prepared by the Contractor as part of the contract, then the Employer should include an appropriate Z clause to provide for it.

2.13 Design of Equipment

The Contractor may be required to carry out design of Equipment, which will include temporary work such as a temporary footbridge and also what would in other contracts be defined as 'Plant'.

Under Clause 23.1, the Contractor submits particulars of the design of an item of Equipment to the Project Manager for acceptance, but only if the Project Manager instructs him to. This is distinct from design of the Works for which the Contractor is required to submit particulars of his design.

A reason for not accepting is that the design of the item of Equipment will not allow the Contractor to Provide the Works in accordance with the Works Information, the Contractor's design which the Project Manager has accepted, or the applicable law.

2.14 Advising on the practical implications of the design of the Works

Under Main Options C, D, E and F, the Contractor has additional obligations in respect of design in that, whether he has design responsibility or not, he is required to advise the Project Manager on the practical implications of the design of the Works.

2.15 Fitness for purpose v reasonable skill and care

In order to consider the design responsibilities of the parties, we should first consider the principles of 'fitness for purpose' and 'reasonable skill and care'.

If one considers a building project as an example, 'fitness for purpose' means producing that building, part of a building, or a product fit for its intended purpose. This is an absolute duty independent of negligence. In the absence of any express terms to the contrary, a Contractor or Subcontractor who has design responsibility will be required to design 'fit for purpose'.

'Reasonable skill and care' means performing to the level of an ordinary competent person exercising a particular skill. The criterion is that it is not possible or reasonable to judge someone's professional ability purely by results; for example, a lawyer whose client is found guilty is not necessarily a poor lawyer, and a surgeon whose patient dies is not necessarily a poor surgeon. The question to ask is 'did that professional apply the correct level of skill and care in carrying out their duties?'

If one then applies the same criterion to a Designer, this is normally achieved by the Designer following accepted practice and complying with statutory requirements, codes of practice, etc, but if a new construction technique is involved the duty can be discharged by taking advice from consultants and warning the Employer of any risks involved. A further qualification on the Designer's liability is the 'state-of-the-art' defence, meaning that a Designer is only expected to design in conformity with the accepted standards of the time.

In the absence of any express terms to the contrary, a Designer will normally be required to design using 'reasonable skill and care'.

Design and build contracts can impose a higher standard than reasonable skill and care, ie fitness for purpose. This obligation resembles a seller's duty to supply goods which are fit for their intended purpose.

The 'default' position within the ECC is that any design carried out by the Contractor must in all respects be fit for purpose. As stated above, this means producing a building, part of a building, or a product fit for its intended purpose. This is an absolute duty independent of negligence.

In the absence of any express terms to the contrary, a construction contract which includes design requires the Contractor to design 'fit for purpose'.

The JCT 2011 Design and Build Contract states that:

> the Contractor shall in respect of any inadequacy in such design have the like liability to the Employer, whether under statute or otherwise, as would an Architect or, as the case may be, other appropriate professional Designer holding himself out as competent to take on work for such design who, acting independently under a separate contract with the Employer, has supplied such design for or in connection with Works to be carried out and completed by a building Contractor who is not the Supplier of the design . . .

This is the JCT version of 'reasonable skill and care'!

Under Option X15, the implied 'fitness for purpose' obligation can be changed to a 'reasonable skill and care' requirement which is effectively a relaxation of the default position.

'Reasonable skill and care' means designing to the level of an ordinary competent person exercising a particular skill. This is normally achieved by the Designer following accepted practice and complying with standards, codes of practice, etc.

Note the 'state-of-the-art' defence that could be applied in the case of a new type of material or a new and as yet untried construction technique, in which case the duty can be discharged by taking advice from consultants and warning the Employer of any risks involved.

In the absence of any express terms to the contrary, a contract for design requires the Designer to design using 'reasonable skill and care'.

The Contractor is not liable for Defects in the Works due to his design so far as he proves that he has used reasonable skill and care to ensure that his design complied with the Works Information and any other required criteria.

If the Contractor corrects a Defect for which he is not liable under this contract, it is a compensation event.

2.16 Dealing with planning permission

The responsibility for obtaining planning permission can lie with the Employer or can be assigned to the Contractor.

In some cases, the Contractor may be responsible for gaining outline planning for the project, however this should be a two-stage process so that if outline planning permission is not obtained the second stage would not proceed.

It is more usual for the Employer to gain outline planning permission and the Contractor to be responsible for gaining full and detailed planning consent.

3 Time

3.1 Introduction

Time, and more specifically starting dates, Completion Dates and Key Dates, the Contractor's programme and other time-related issues are covered in the ECC by Clauses 30.1 to 36.2 with further references within the Main Option clauses.

3.2 General principles of planning and programme management

A successful construction project is considered by Employers and Consultants to be one which has been completed on time, to the required quality and also within budget; and by Contractors and Subcontractors as one which has been completed on time, to the required quality and making them a profit! Clearly, apart from their different perspectives on financial outcome, the objectives of both contracting parties are well matched.

When considering Completion on time, various surveys carried out in recent years have shown that only about 33 per cent of contracts are completed within the contract period, one fifth overrun by 40 per cent and a significant number overrun by more than 80 per cent!

There are many factors which can cause delays:

- delays/errors in design, often exacerbated by poor communications between design and construction teams;
- inadequate Site investigations, Employers often believing that Contractors should be responsible for assessing ground conditions based on poor and inadequate information provided by them;
- inability to give timely access or possession;
- non-availability of resources, by all parties;
- poor performance of parties;
- Employer changes, compounded by the number or timing of these instructions;
- weather and other seasonal problems;
- insolvency of a party;
- unrealistic construction period set by Employer.

Clearly, matters need to be improved by the parties and between the parties, but many of these factors can be avoided or at least reduced by the parties engaging in proper and effective planning and programming of the Works aligned to effective communication between the parties.

3.3 Planning and programming

The terms 'planning' and 'programming' are used interchangeably; however, their meanings are a little different.

What is planning?

Planning may be defined as 'the deliberate consideration of all the circumstances concerned with a project in order to evolve the best method of achieving a stated objective'. This involves defining the project, setting the overall duration and identifying risks.

Planning is a vital element in any construction process without which it is impossible to envisage the successful conclusion of any project.

Planning considers:

- what to do;
- when to do it;
- how to do it;
- what resources are required to do it;
- who is going to do it;
- how long it will take to do it.

The motto is:

PLAN YOUR WORK . . . AND WORK TO YOUR PLAN!

Planning may be considered on a long- or short-term basis; long-term planning includes the development of a master plan for the whole project, whilst short-term planning is a much more detailed operation, which considers only the next three to four weeks of operations, or a particular element of the work. It is essential that each short-term plan should overlap the other by one to two weeks to allow for past and future activities to be considered.

The planning process considers each operation in turn, analysing the order and sequence of each operation, its dependency on previous operations, and the Contractor's and Employer's resources to carry it out and the availability of those resources.

It is easier to track progress of operations using short-term, rather than long-term planning.

There are three planning stages:

Pre-tender planning

This covers the planning considerations during the preparation of a cost estimate and its conversion into a submitted tender Price.

The process enables the Contractor to:

- establish a realistic construction period on which the tender may be based – the Employer may set the starting date and the Completion Date for the work or may set the starting date and invite the tendering Contractors to propose a Completion Date;
- consider the most effective economical construction methods, using the most effective resources;
- assist the build-up and pricing of contract establishment charges and Equipment expenditure.

This is usually carried out in broad outline format as the Contractor may not be successful in his tender.

Typical information which may be gathered and used at the tender planning stage includes:

- a brief description of the project and its location, details of the Employer, and the various consultants, key companies, third parties and individuals;
- details of the local planners and any planning issues or restrictions;
- access to Site for delivery and collection of Equipment, Plant and Materials and also removal of waste;
- details of existing services and the location of new and existing connection points, and details of the relevant service providers;
- general details of the geography and topography of the area, past use, ground investigations, groundwater levels and other local knowledge;
- location of tips and Suppliers including quarries;
- availability of labour, Equipment, Plant and Materials;
- details of possible Subcontractors;
- typical weather conditions.

This information is then collected together with information given to the Contractor at tender stage to develop a pre-tender plan. Time-based elements can then be transferred to a pre-tender programme, usually in a bar chart format which together with the method statements then aids the estimator in preparing the tender and assessing the need for establishment charges and major multi-use Equipment such as craneage, scaffold and skips.

Employers should instruct Contractors to submit at least outline programmes with their tenders, though it must be appreciated that as the Contractor may not be successful with his tender it will serve only as an indicative outline of how the Contractor would carry out the Works should he be successful.

The ECC provides as an optional statement for the Contractor to submit a first programme for acceptance within a specified number of weeks of the Contract Date.

Pre-contract planning

This covers the planning considerations once the contract has been awarded. It is essential that this planning stage takes place prior to commencing work on the Site.

The process enables the Contractor to:

- establish a clear strategy for carrying out the Works;
- comply with contract conditions and requirements;
- establish a construction sequence on which the master programme may be based;
- identify Key Dates by which parts of the work are required to be completed or the Employer is to provide information or resources to enable the Contractor to complete work to the programme requirements;
- enable the assessment of contract budgets and cash flow forecasts;
- schedule Key Dates for procurement of key Materials and Subcontractor tendering, placing orders and commencement;
- schedule pre-contract meetings.

Project planning

This covers the planning which is required to be implemented in order to maintain control and ensure that the project is completed on time and within the cost limits established at the tender stage.

The process enables the Contractor to:

- monitor and update the master programme on a regular basis;
- optimise and continually review resources;
- keep progress under review and report on variances.

What is programming?

Programming is drawing all the planning considerations together into a presentable graphical format such as a bar chart and supporting information and tables which show the various operations, their durations, relationship between the operations, critical path and float, and the required resources so that it can be used as a project reporting and control mechanism, and form the basis of progress reporting to the Employer.

Programming is an essential part of any construction contract, but whilst other contracts often pay only a passing interest to the programme, the NEC3 family sees it as a vital tool from which to:

- at tender stage, assess the Contractor's ability to carry out the project;
- measure and report the Contractor's actual progress against that planned;
- assess the effect of compensation events upon Completion and Price;
- ensure the Project Manager takes appropriate action when required.

3.4 Preparation of a programme

The process of writing a programme normally consists of three stages:

1. *Identifying the various operations to be carried out*
   The operations may initially be written as a list, identifying each operation in chronological order. That order may depend on natural construction sequence, constraints to access and resource availability.

2. *Considering the duration of each operation*
   The duration of each operation can only be a forecast rather than an exact science as various issues have to be considered such as length of working days, differing productivity outputs, changes, resource availability, working conditions, changing weather and Equipment breakdowns which can affect forecast durations. The Contractor will then make appropriate adjustments as the work progresses.

3. *Considering the relationship between each operation*
   This will involve inserting links between operations which have a dependency on each other, for example 'start to start', 'finish to start', 'finish to finish'. From this can be established what the logical path is in order to execute the various operations, operations that need to start at the same time, what operations have to be completed before another operation can start, and what operations must finish at the same time.

Figure 3.1 shows a simple programme for the construction of two storage tanks. One can interpret from the programme that there is an initial period for

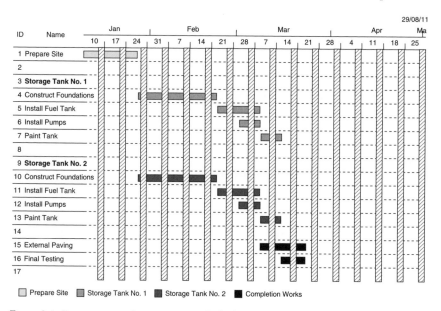

Figure 3.1 Programme with operations unlinked

preparing the Site, the construction of the foundations to both tanks then commencing once the Site has been prepared.

The construction of the tank foundations, the installation of the tanks themselves, and the installation of the pumps and tank painting for both tanks then progresses concurrently. Whilst the tanks are being painted, the external paving around the tanks can then progress, then the final operation, final testing, will progress with Completion of this operation coinciding with Completion of the external paving, and thus Completion of the Works.

The problem is that without any links, the dependency of the operations upon each other, whilst implied from their positioning within the programme, is not definitive; therefore if, for example, the construction of the foundations to Tank No. 1 is delayed, either by the Contractor or the Employer, the effect on other operations can be inferred but is not stated definitively.

Figure 3.2 includes links which show the logic and dependency of each operation, therefore, if the construction of the foundations to Tank No. 1 is delayed, it is on the critical path and therefore the delay to the foundations will have a direct delay effect on all the following operations.

The term 'programme' is not defined in law or within contracts, the dictionary definition normally stating that a programme is 'a plan or schedule of activities to be carried out'. The ECC does not prescribe the form the programme submitted for acceptance must take, it would normally be a bar chart programme.

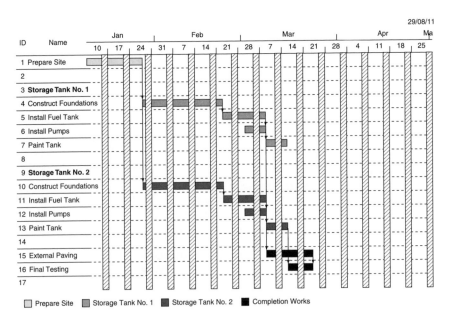

Figure 3.2 Programme with operations linked

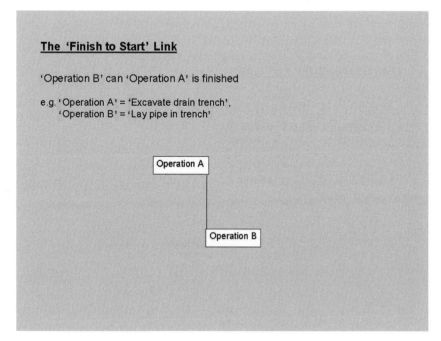

The 'Start to Start' Link

'Operation B' can start at the same time as 'Operation A'

e.g. 'Operation A' = 'Excavate drain trench',
   'Operation B = 'Remove excavated material to spoil heaps'

Operation A

Operation B

Figure 3.3a The 'Start to Start' Link

The 'Finish to Start' Link

'Operation B' can 'Operation A' is finished

e.g. 'Operation A' = 'Excavate drain trench',
   'Operation B' = 'Lay pipe in trench'

Operation A

Operation B

Figure 3.3b The 'Finish to Start' Link

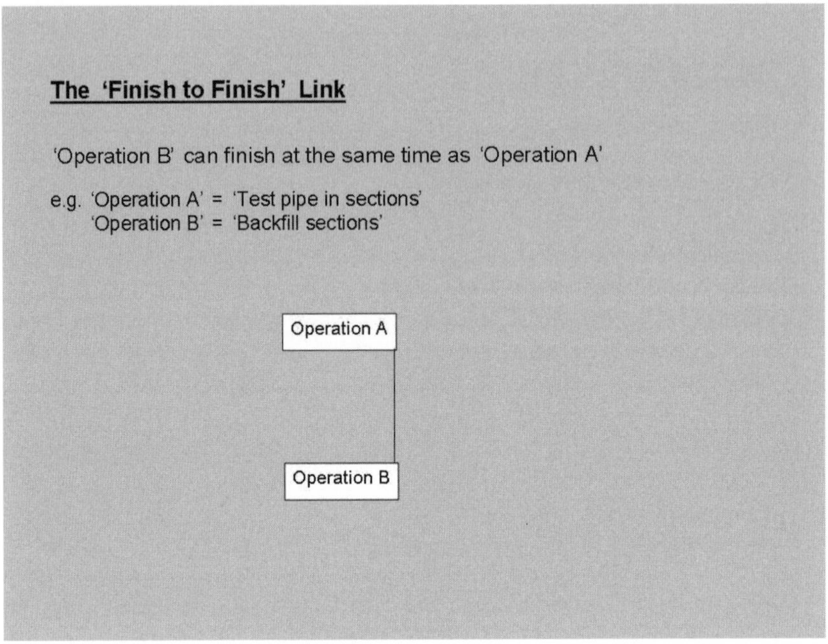

The 'Finish to Finish' Link

'Operation B' can finish at the same time as 'Operation A'

e.g. 'Operation A' = 'Test pipe in sections'
     'Operation B' = 'Backfill sections'

Operation A

Operation B

Figure 3.3c The 'Finish to Finish' Link

The 'Start to Start' Lag Link

'Operation B' can start x days after 'Operation A' starts

e.g. 'Operation A' = 'Pour concrete'
     'Operation B' = 'Remove formwork'

Operation A

Operation B

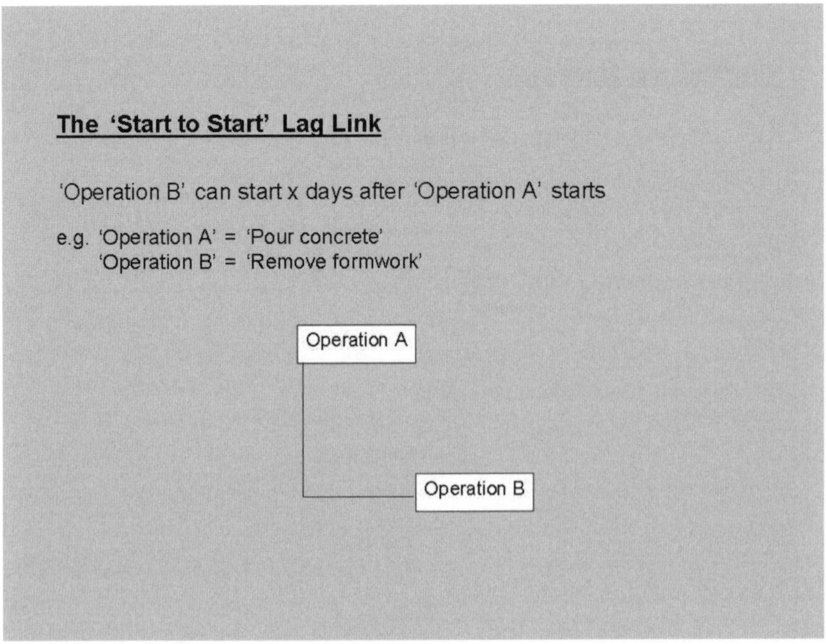

Figure 3.3d The 'Start to Start Lag' Link

3.5 Types of programme

Before considering in detail the requirements of the ECC, it is worth reviewing types of programme in common use.

(i) Bar charts

Any programme, at any stage of a tender or project, is normally written as a bar chart or Gantt chart (Henry Laurence Gantt (1861–1919)) showing operations which may occur either on or off Site including their sequence and duration, the start and finish of the bar depicting the start and finish date of the operation. Linked bar charts as described previously are the usual form of programme in common use on construction contracts. They are often supplemented by other information.

Advantages of bar charts

- The simple format allows them to be readily constructed, they are user-friendly and understood by most practitioners.
- They illustrate the specific activities, their durations and during the progress of the work, actual versus planned progress can be clearly shown.
- They can be used at all stages of the planning process through pre-tender, pre-contract and contract planning. They assist in showing the relationship between the pre-tender programme, the master programme and short-term programmes.
- They clearly relate to the construction sequence, the use of linking on bar charts aids the overlapping of related activities and can enable the critical path to be identified.
- They are easily updated, at weekly and monthly time intervals for review purposes and progress reports.
- Key milestone symbols may be introduced to highlight critical dates with regard to key contract stages and information requirements.
- Resources may be readily related to the rate of working – labour histograms, value forecasts (value/time), and cumulative labour and Plant forecasts in the form of project budgets.
- The programme forms the basis of financial forecasting for both the Employer and the Contractor as a cash flow forecast can be prepared by pricing the programme.

Disadvantages of bar charts

- Without the use of links they cannot properly depict the relationship and dependency between activities, for example where an activity must be wholly or partially complete before the next can start.
- Complex interrelationships cannot always be clearly shown.
- They may give too simple a picture, which may be misleading.

(ii) Elemental trend analysis (line of balance) charts

The elemental trend analysis (line of balance) charts, as with many programming methods, originated from the manufacturing industry and, whilst little known or used, are beneficial where there is repetition of work either in number of units such as housing, or in repetitive floors in constructing or refurbishing multi-storey buildings. The unit or floor numbers are shown on the vertical y-axis, and the timeline of the project on the horizontal x-axis.

Advantages of elemental trend analysis charts

- They are very suitable for repetitive work and strict trade sequencing as the flow of work can be clearly demonstrated.
- They are more sophisticated than bar charts and simpler than network diagrams.
- They clearly show the rate of progress of each trade.

Disadvantages of elemental trend analysis charts

- They are totally unsuitable for non-repetitive projects and largely unsuitable for limited repetition.
- They are not easily understood, user-friendly, or readily understood by all practitioners.

(iii) Project network techniques (precedence diagrams, network analysis, critical path analysis)

Project network techniques comprise a number of methods but all are related to the dependency and precedence of one activity over another.

Networks may be represented as arrow diagrams where activities are represented by arrows and events (start and finishes) by 'nodes'. Activities take time, and are represented as months, weeks, days or even hours, but events do not.

As stated earlier, the bar chart is a clear and simple method of plotting activities, but it has a clear disadvantage in that the relationship between activities cannot be properly represented. With networks, the relationship is immediately apparent and when a delay occurs specifically affected activities will stand out.

A project network illustrates the relationships between activities (or tasks) in the project, showing the activities either on arrows or nodes.

Preparing network diagrams

The first operation as with any method is to compile a list of activities to be carried out.

One must then consider against each activity:

- Which activities must be completed before this activity can start?
- Which activities cannot start until this one is completed?

NB One should always consider when an activity can be done rather than when one thinks it will be done.

Advantages of network diagrams

- They are extremely effective when one is considering highly critical activities with strict dependency of several activities upon each other.
- Float and critical path can be calculated fairly easily from such diagrams.

Disadvantages of network diagrams

- They require a great deal of time and effort to produce.
- They are not practical for 'normal' construction activities, which do not require highly detailed planning and monitoring of progress.
- They do not provide a ready format for financial reporting.

Most major projects are managed using software which combines high-quality presentation with ease of revision and updating. This software can also link the master programme to short-term programmes and also resource schedules and cash flow forecasts, however one must never forget the power of pen and paper.

3.6 The traditional approach of construction contracts

The value of a fully comprehensive construction programme is often misunderstood and undervalued by construction practitioners, this is often exacerbated by the minimal express requirement for a programme in most standard forms of construction contract.

Various commonly used forms merely state that the Contractor 'submits his master programme for the execution of the Works', 'states the order in which he proposes to carry out the Works' or 'shall submit a detailed time programme . . . and shall submit a revised programme whenever the previous programme is inconsistent with actual progress or with the Contractor's obligations', often without going into any detail as to the required content and structure of the programme, the timing of the submission, the timing of the submission of any revised programme or any formal acceptance of the Contractor's programme.

The danger is that by understating the requirement for a programme within the contract, the true value of a programme is then misunderstood by the parties to that contract, and the programme submission and any acceptance can then descend into a 'box ticking' exercise.

Clearly, although the programme requirement within these contracts is not spelt out in detail, it is advisable on any project for the Employer to require the

Contractor to submit a programme, even though there may be no contractual requirement to do so, and it is also advisable for the Contractor to submit a detailed programme even if there is no contractual requirement to do so. The submitted programme should also identify the requirements to be met by the Employer.

The programme must be a dynamic tool which addresses all aspects of the project including design, information and other requirements, and the procurement process and lead in times for Subcontractors and Suppliers in addition to the construction operations. It must be flexible, and reviewed and, if necessary, revised on a regular basis as the work progresses and various events and changes come to light including the adjustments to float and time risk allowances.

3.7 The ECC approach

The ECC takes a different approach to other contracts in dealing with the programme by clearly stating in detail what the Contractor is required to include in the programme he submits, with sanctions if he fails to do so, and also requiring that the Project Manager either accepts the programme, or gives his reasons for not accepting it.

The clear programme requirement enables it to be used to assess:

- the Contractor's ability to carry out the project in terms of the intended resources and within the contractual time frame;
- whether all of the contract needs and requirements have been considered;
- the Contractor's actual progress against that planned;
- the effect of compensation events upon Completion and Price.

Dates in the contract

There are a number of dates in the ECC which require some definition.

(i) Contract date

The Contract Date is the date the contract came into existence, however this is achieved, for example by the Employer issuing a letter of acceptance to the Contractor. Note that there is no pro forma contract agreement within the contract, the principle being that parties will enter into contracts in various ways, using their own pro formas, standard letters or exchange of correspondence.

(ii) Starting date

In contrast with most forms of contract, the period of time for Completion of the Works is not stated. Instead, the starting date is decided and defined by the Employer in the Contract Data together with the Completion Date. This has the advantage that tenderers know exactly when the Works are required to start and to complete.

The starting date and the access dates are stated in the Contract Data. The Contractor can start work from the starting date, but does not have access to the Site and cannot physically start on Site before the first access date. He also carries risks from the starting date and so has to provide the insurances in the contract prior to that date.

(iii) Access date(s)

The date(s), on which the Contractor may have access to and use of various parts of the Site, are stated in the Contract Data. Whilst the Contractor can start work on the contract from the starting date, he cannot start work on the Site until the first access date. If the Employer has stated different access dates for different parts of the Site, then the Contractor must programme his activities according to the dates when he will be given access.

The Contractor may not require access on the dates stated in the Contract Data, in which case he should show on his programme the dates when he requires access. The Employer is then required to give access to parts of the Site by the later of the access date in Contract Data Part 1, and the date for access shown on the Accepted Programme (Clause 33.1).

(iv) Completion date

The Contractor is obliged to complete the Works on or before the Completion Date as stated in the Contract Data or as may be revised in accordance with the contract.

The Project Manager decides the date of Completion, and certifies within one week of Completion (Clause 30.2). Note that the Project Manager is not deciding when the Contractor has to be complete by, but when he has completed.

Failure to complete on time constitutes a breach of the Contractor's obligation. If the Secondary Option X7 (Delay damages) is included, the Employer is then entitled to pre-defined damages for that delay. If X7 is not included then unliquidated damages will apply.

Whilst many contracts use the terms 'Practical Completion' and 'Substantial Completion', the ECC does not, as these terms are often subject to various interpretations. The ECC is more objective in defining Completion under Clause 11.2(2) as when the Contractor has:

- *done all that the Works Information requires him to do by the Completion Date;*
  This first bullet point requires the Contractor to comply with the requirements of the Works Information and supersedes the traditional term 'Substantial Completion'. This requirement may include not only physical Completion of the Works, but also submission of 'as-built' drawings, maintenance manuals, training requirements, successful testing requirements, etc. By

Contract:	**COMPLETION CERTIFICATE**
Contract No:	**C/C No.......................................**

Completion achieved on: .... (Date)

The Completion Date is: .... (Date)

The *Defects Date* is: .... (Date)

The *Defects on the attached schedule are to be*
corrected within the Defect Correction Period
which ends on: .... (Date)

The *outstanding Works on the attached schedule are to be*
completed by: .... (Date)

Comments:

Signed: ....(Project Manager)   Date: ....

Copied to:

Contractor ☐ Project Manager ☐ Supervisor ☐ File ☐ Other .... ☐

Figure 3.4 Sample Completion Certificate

including these requirements within the Works Information, Completion has not been achieved until they are completed.

- *corrected notified Defects which would have prevented the Employer from using the Works and Others from doing their work.*

The Works may contain Defects, though these are not significant enough to prevent the Employer from practically and safely using the Works. Completion is certified with the requirement that the Contractor corrects the Defect before the end of the Defect Correction Period following Completion.

There is also a fall back within Clause 11.2(2) in that if the work which the Contractor is to do by the Completion Date is not stated in the Works Information, Completion is when the Contractor has done all the work necessary for the Employer to use the Works and for Others to do their work. This broadly aligns with the traditional terms 'Practical Completion' and 'Substantial Completion', where the Employer is able to take beneficial occupation of the Works and use them as intended.

The Project Manager is responsible for certifying Completion, as defined in Clause 11.2(2), within one week of Completion. Normally, the Contractor will request the certificate as soon as he considers he is entitled to it, but such a request is not essential.

(v) Key Date

This is defined in Clause 11.2(9) as the date by which a stated Condition has to be met. It is an optional clause allowing the Employer, should he wish, to include dates when the Contractor must complete certain items of work, possibly to allow others to carry out other work. If used, the Condition and the applicable Key Date are defined by the Employer in Contract Data Part 1.

'Key Dates' must be differentiated from 'Completion' or 'Sectional Completion', in that with Key Dates the Employer does not take over the Works, whereas with Completion and Sectional Completion, he does. In this respect there are no pre-determined delay damages applied to the Contractor who does not meet a Condition by a Key Date.

However, if the Project Manager decides that the Contractor has not met the Condition stated by the Key Date, and the Employer incurs additional cost on the same project as a result of that failure, then the Contractor will be liable to pay that amount (Clause 25.3). By referring to 'the same project', the Employer cannot claim costs incurred on another project as a result of the failure of the Contractor to meet a Key Date. This cost is assessed by the Project Manager within four weeks of when the Contractor actually does meet the Condition for the Key Date.

Example

The Contractor is building a new outpatients department for a local hospital. The Employer wishes to have new x-ray machinery installed by a specialist who he will employ directly and who will install the machinery as the building work progresses. In order to do that the Contractor has to complete the part of the building that will house the machinery and install the necessary electrical power facilities so that the machinery can be tested prior to Completion of the project.

The necessary work to be carried out by the Contractor in readiness for the specialist to carry out his part will be the Condition and this will have a Key Date attached to it.

If the Contractor then fails to meet the Condition by the Key Date and the Employer incurs a cost in having to postpone the installation of the x-ray machinery, then this cost is paid by the Contractor. The cost must be incurred on the same project, so, for example, if the late installation of the x-ray machinery incurs a cost on another project, then this is not paid by the Contractor.

The ECC has quite comprehensive and specific requirements for a programme to be submitted by the Contractor, these requirements being defined by 17 bullet points within Clause 31.2.

Some practitioners say that the requirement within Clause 31.2 is very onerous and in some cases excessive; however, it is critical that the programme shows in a clear and transparent fashion what the Contractor is planning to do, when and how long each operation will last, and what he needs for the Employer and others to be able to comply with it.

There is nothing within the requirements of Clause 31.2 that a competent Contractor would not ordinarily include within a professionally constructed programme submitted for the benefit of himself and the receiving party. Whether and how he chooses to show the information could be another matter!

There are two alternatives for the Contractor's submission of his first programme for acceptance:

1. He may submit, or be required to submit, a programme with his tender in which case it is referenced by the Contractor in Contract Data Part 2. Note that under Clause 11.1(1) 'the Accepted Programme is the programme identified in the Contract Data or is the latest programme accepted by the Project Manager'. Therefore, any programme referred to in Contract Data Part 2 automatically becomes the first Accepted Programme even though it may not necessarily comply fully with the requirements of Clause 31.2.

2. If a programme is not identified in Contract Data Part 2, the Contractor submits a first programme to the Project Manager for acceptance within the period of time after the Contract Date, this period of time being stated in Contract Data Part 1.

   There is no default position, found in some other contracts, where no response to a programme submission is deemed as acceptance of it.

Contract:	**CONTRACTOR SUBMISSION**
Contract No:	C/S No..

Section A:	Submission of:
To : Project Manager/Supervisor	☐ **Drawings**
	☐ **Programme**
	☐ **Test Results**
	☐ **Other**

The following information is transmitted for your acceptance:-

Description	Date	No./Rev.	Copies

Response reqd. by ....

Signed: (Contractor) ....   Date ....

Section B: Reply
To: Contractor
The submission is returned:-

☐ **Accepted**   ☐ **Accepted as noted**
☐ **To revise and resubmit**   ☐ **Rejected as noted**

Notes:

Signed: (Project Manager) ....   Date:...............

Copied to:

Contractor ☐ Project Manager ☐ Supervisor ☐ File ☐ Other .... ☐

Figure 3.5 Sample Contractor submission

Such is the importance of a programme within the ECC that if no programme is identified in Contract Data Part 2, one-quarter of the Price for Work Done to Date is retained (Clause 50.3) until the Contractor has submitted the first programme showing the information the contract requires. Note that a Contractor who submits a first programme which shows the required information, but the

Project Manager does not accept it, would not be liable for withholding of payment under this clause. The clause relates to the Contractor's failure to submit a programme showing the information the contract requires, not the acceptance of it.

Clause 31.2

The Contractor shows on each programme which he submits for acceptance:

- *the starting date, access dates, Key Dates and Completion Date*
  These dates are stated by the Employer in Contract Data Part 1. With regard to Completion Date, if the Employer has decided the Completion Date for the whole of the Works then it is inserted into Contract Data Part 1. Alternatively, if the Contractor is to decide the Completion Date he inserts this into Contract Data Part 2 as part of his tender.

- *planned Completion*
  This is the date the Contractor *plans* to complete the Works; this is different to the date by which the Contractor is *required* to complete the Works as stated in Contract Data Part 1, or as may be revised in accordance with the contract. Although the Contractor can complete any time between his planned Completion Date and the Completion Date, failure to complete on or before the Completion Date constitutes a breach of the Contractor's obligation and if Option X7 has been selected, the Contractor is liable to pay the Employer delay damages.

  The difference between the date when the Contractor plans to complete and when he is required to complete is float within the programme. This float, often referred to as 'terminal' float belongs to the Contractor and is attached to the whole programme. If a compensation event occurs, a delay to the Completion Date is assessed as the length of time that planned Completion is later than planned Completion as shown on the Accepted Programme (Clause 63.3).

  It is vital that as planned Completion can change on a regular basis, the Contractor shows a realistic planned Completion in each revised programme submitted for acceptance, but also that the Contractor and Project Manager comply with the time scales in the contract for assessing and implementing compensation events as it is not unusual for planned Completion to be shown later than the Completion Date leading one to believe the Contractor is in delay, yet the reason for this is either that the time scale in the contract is being adhered to but is giving a false picture as a programme is submitted whilst the Contractor is still preparing his quotation for a compensation event, or that compensation events have not been assessed and implemented in accordance with the time scales stated in the contract.

- *the order and timing of the operations which the Contractor plans to do in order to Provide the Works*

This shows the sequence, duration and dates of the various operations the Contractor plans to do, and should include dependencies and links between the various operations. It should also show sequence, duration and dates in the procurement process, finalisation of information, placing orders, etc. It should also show work to be carried out off Site.

- *the order and timing of the work of the Employer and Others as last agreed with them by the Contractor or, if not so agreed, as stated in the Works Information*
  The Contractor will state what he requires the Employer and Others to do in order that the Accepted Programme can be met. These dates may already have been agreed with the Contractor, but may also have been inserted into the Works Information by the Employer. 'Others' are parties outside the contract, ie not the Employer, Project Manager, Supervisor, Adjudicator, the Contractor or a Subcontractor or Supplier to the Contractor. Examples of 'Others' would be planning authorities, utilities companies, etc.
  This requirement is often provided in the form of an Information Required spreadsheet identifying what information is required and the latest date for providing it, failing which a compensation event can occur (Clause 60.1(5)). It is important to recognise that the programme is not just the bar chart but may include spreadsheets, schedules and graphs.

- *the dates when the Contractor plans to meet each Condition stated for the Key Dates and to complete other work needed to allow the Employer and Others to do their work*
  If there are Key Dates in the contract (Clause 11.2(9)), then the Contractor must show how and when he intends to meet each Condition by these Key Dates. There may be provision for other Contractors directly employed by the Employer to carry out work, for example a fit out Contractor.

Provisions for:

- *float*
  Float is any spare time within the Contractor's programme, after time risk allowances have been included, and represents the amount of time that operations may be delayed without delaying following operations (free float) and/or planned Completion (total float). It can also represent the time between when the Contractor plans to complete the Works and when the contract requires him to complete (terminal float).
  Float absorbs to a certain extent the Contractor's own delays or the delays caused by a compensation event, thereby lessening or avoiding any delay to planned Completion. In effect, no delay arises unless float on the relevant and critical operations reduces to below zero.
  Programming is never an exact science, so float gives some flexibility to the Contractor in respect of incorrect forecasts or his own inefficiencies. As the work progresses the float will change as output rates change, Contractor's risk events take place and also compensation events arise.
  The general belief, certainly amongst many Contractors, is that float

belongs to them, as they wrote the programme, and therefore have the right to work to that programme and use any float that it contains. Conversely, Employers believe that if it is free time then they have the right to use it, but it actually depends where the float is and what it is for.

In principle, float other than terminal float or time risk allowances is a shared resource; it is spare time, it belongs to the project, and therefore may be used by whichever party needs it first.

There are three primary types of float:

1. the amount of time that an operation can be delayed before it delays the earliest start of the following operation (free float);
2. the amount of time that an operation can be delayed before it delays the earliest Completion of the Works (total float);
3. the amount of time between planned Completion and the Completion Date (terminal float).

Generally, with other contracts, if the Contractor shows that he plans to complete a project early, then he is prevented from completing as early as he planned, but he still completes before the contract Completion Date, then although there may be an entitlement to an extension of time under the contract, for example exceptionally adverse weather, none will be awarded as there is no delay to contract Completion, but he may have a right to financial recovery subject to him proving loss and/or expense.

The ECC deals with this in a different way. If the Contractor shows on his programme planned Completion earlier than the Completion Date and he is prevented from completing by the planned Completion Date by a compensation event, then, when assessing the compensation event, Clause 63.3 states 'a delay to the Completion Date is assessed as the length of time that, due to a compensation event, planned Completion is later than planned Completion as shown on the Accepted Programme'; therefore any terminal float is retained by the Contractor, the period of delay being added to the Completion Date to determine the change to the Completion Date.

Any delay to planned Completion due to a compensation event therefore results in the same delay to the Completion Date. Therefore, the extension is granted on the basis of time the Contractor is delayed, ie entitlement, not on how long he needs to achieve the current Completion Date.

Example

The Completion Date is 18 November 2011, but the Contractor is planning to complete by 4 November 2011, the programme showing the last operation (roof covering) being completed on that date. The Project Manager has accepted the Contractor's programme.

> At the beginning of October 2011, the Project Manager gives an instruction to change the specification for the roof from concrete tiles to slates. This is a change to the Works Information and therefore a compensation event under Clause 60.1(1). The Contractor prepares his quotation and finds that he cannot get the slates delivered until week commencing 7 November 2011; therefore the roof cannot be completed until 11 November 2011 and the Contractor then shows in his quotation a delay to planned Completion of one week.
>
> As Clause 63.3 states that 'a delay to the Completion Date is assessed as the length of time that, due to a compensation event, planned Completion is later than planned Completion as shown on the Accepted Programme', assuming the Project Manager accepts the quotation, planned Completion becomes 11 November 2011 and the Completion Date becomes 25 November 2011.

- *time risk allowances*

Time risk allowances are essentially a form of float and are included within the duration of a specific operation. Essentially, a time risk allowance is the difference between what is the most 'optimistic' (shortest) duration to complete an operation and what is a 'realistic' (expected) duration bearing in mind the risks that the Contractor may face in completing that operation. It is important that the Contractor's staff are fully informed as to how the time risk allowances are being presented, so there is no misunderstanding of how the Works have been programmed and priced.

The Contractor's time risk allowances are to be shown on his programme but there are no requirements in the contract as to how these allowances should be shown, three possible methods being:

1. show a single bar for an operation, within which the time risk allowance is shown – this is difficult to do with some planning software;
2. show a separate bar which represents the time risk allowance for a specific operation – this, however, adds to the lines on the programme;
3. show a separate column identifying the time risk allowance included within the total duration of the operation.

Clause 63.6 states that the assessment of the effect of a compensation event includes risk allowances for cost and time for matters which have a significant chance of occurring and are at the Contractor's risk. It follows that they should be retained in the assessment of any delay to planned Completion due to the effect of a compensation event. These allowances are owned by the Contractor as part of his realistic planning to cover his risks in pricing his tender and also in managing changes. They should be either clearly identified as such in the programme or included in the time periods allocated to specific activities.

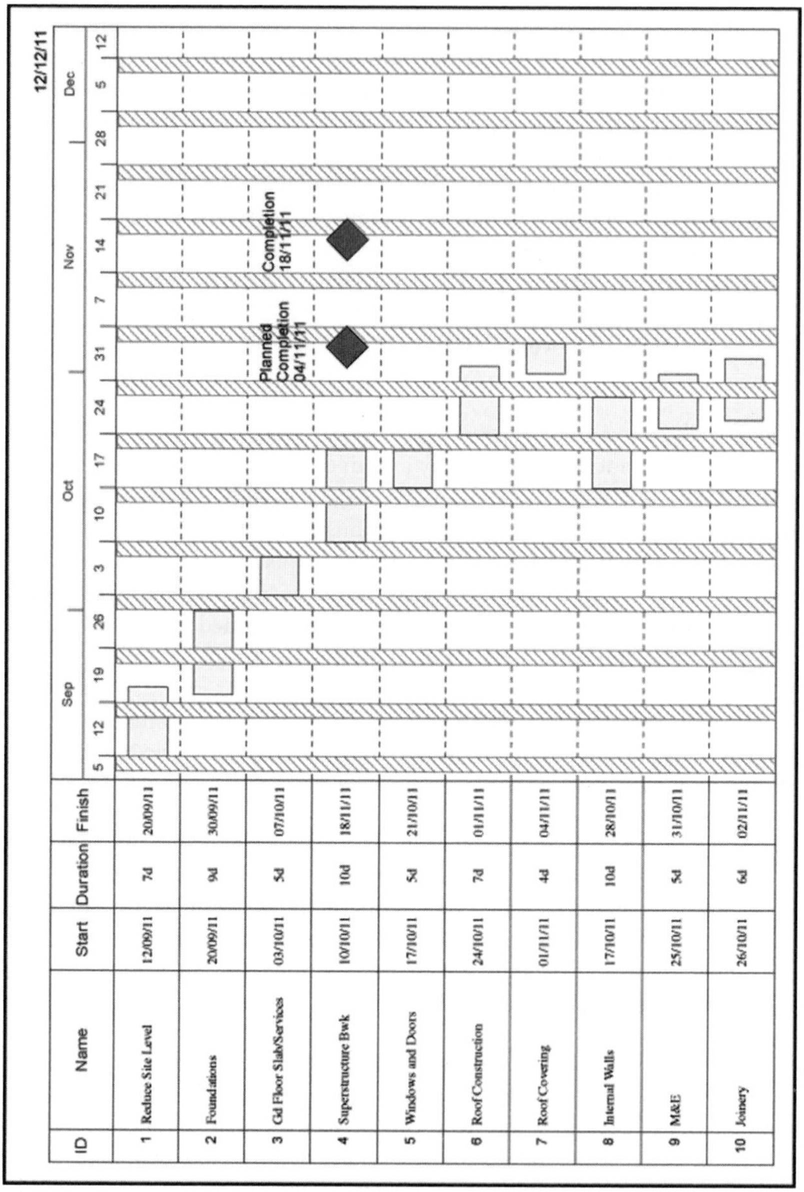

Figure 3.6 Terminal float example

Example

The Contractor has an operation in his programme for 'piling'. He believes that if the ground conditions are reasonable, the weather is reasonable, and there are no mechanical problems with the Equipment, then the piling operation will take him six weeks to complete.

However, his knowledge and experience tells him that rarely is everything reasonable and almost inevitably he will encounter some delay, so a realistic time scale would be eight weeks. He should then show an eight-week duration for the piling operation, within which two weeks are his time risk allowance.

It is important to recognise that time risk allowances belong to the Contractor and are retained in the assessment of any delay to planned Completion due to a compensation event.

- *health and safety requirements*
  This will include time taken in inducting new staff and operatives.

- *the procedures set out in this contract*
  This can include time scales for acceptance of design, etc.

- *the dates when, in order to Provide the Works in accordance with his programme, the Contractor will need*

  - *access to a part of the Site if later than its access date;*
  - *acceptances;*
  - *Plant and Materials and other things to be provided by the Employer;*
  - *information from Others.*

Generally, the Contractor should be given access to parts of the Site, acceptance of his design, Subcontractors and programme in a timely manner to allow him to Provide the Works. However, in respect of access, the Contractor may be in delay and may therefore not require access on the date shown in the contract, there may also be certain operations for which there may be a long lead in time, which the Employer or Others may not appreciate and therefore the Contractor can include the requirement for acceptances and information at an early date.

This information may be highlighted as a milestone on the bar chart itself, or attached to it in the form of information required schedules, resource schedules, etc.

Anything submitted by the Contractor as part of his programme must clearly and simply show the relevant operations for which access, Plant and Materials or information are required, and the implications if they are not provided.

It is important that the Contractor and the Project Manager then manage the process by raising any concerns in respect of their provision as early warnings.

- *for each operation, a statement of how the Contractor plans to do the work identifying the principal Equipment and other resources which he plans to use*
  The Contractor is required to provide a statement of how he intends to carry out each and every item on each programme he submits for acceptance.

  Whilst accepting that the contract requires this information, this can be a somewhat idealistic clause which requires the Contractor to comply with the daunting and in many cases somewhat unreasonable requirement to produce all the information in his first programme submitted for acceptance. Particularly on a large complex project with a long duration, the Contractor may not yet know how he will carry out each operation and what resources he will use. He also may not have appointed a Subcontractor for a part of the Works so will not have method statements from the company who is actually carrying out the Works.

  It is also a significant task for the Project Manager to accept all the information the Contractor has submitted, so it may be appropriate in this respect for the Contractor and the Project Manager to agree a rolling programme of submission of method statements as the work progresses and the methods and resources may become clearer, perhaps each submission covering the next three or four months' operations.

  Method statements have become increasingly important in the construction industry, particularly as health and safety legislation makes the parties more responsible for stating the methodology they are going to use in carrying out the Works, and how they have provided for the various risks involved. For a programme to be meaningful it should always be linked to method statements so that the methodology of carrying out the Works and also the resources the Contractor intends to use are clearly stated.

  The contract does not specify the form these statements should take and in many cases a Contractor will spend a great deal of time and paper producing detailed statements, which can be almost meaningless. They may be presented as a written statement or in tabular form but they should clearly show how the Contractor intends to do the work, the main headings being:

  - title of each operation;
  - description of method;
  - quantities (where relevant);
  - principal Equipment to be used with outputs and durations;
  - labour type and gang sizes with outputs and durations;
  - Plant to be installed;
  - Materials to be used.

NB Cost is not normally included in method statements.

  When preparing method statements, the Contractor should always consider and show:

- what is to be done;
- how it is going to be done;
- who is going to do it;
- where it is to be done;
- when it is to be done.

This information will then initially allow the Project Manager to consider whether the Contractor is providing sufficient resources to Provide the Works within the time available and also safely, but also as the programme is revised and methods of construction and resources used may change the Project Manager is then kept up to date as to the Contractor's intentions.

This information will also prove invaluable when assessing compensation events.

- *other information which the Works Information requires the Contractor to show on a programme submitted for acceptance*
  Again, this shows that the Works Information is not just the drawings and specifications, but information on all that the Contractor is required to do in Providing the Works.

  From these requirements one can see that the programme under an ECC contract is not just the bar chart, but a collection of documents including method statements, risk assessments, resource analyses, etc. All of these are submitted by the Contractor as his programme, and all of these must be considered by the Project Manager in determining whether to accept the programme. Extreme care should be taken to require only documentation which has a purpose and use in the administration.

3.8 Activity schedule-based Options A and C

Clause 54.2 of Options A and C requires that if the Contractor changes his method of working so that the activities on the activity schedule do not relate to the operations on the Accepted Programme, he submits a revision of the activity schedule to the Project Manager for acceptance.

Under Clause 54.3, a reason for not accepting a revision to the activity schedule is that:

- it does not comply with the Accepted Programme;
- any changed Prices are not reasonably distributed between the activities; or
- the Total of the Prices is changed.

3.9 The Project Manager's response

Clause 31.3

Within two weeks of the Contractor submitting a programme to him for acceptance, the Project Manager either accepts the programme or notifies the Contractor of his reasons for not accepting it.

A reason for not accepting a programme is that:

- *the Contractor's plans which it shows are not practicable*
  The Contractor could be planning to use a particular working method or piece of Equipment that the Project Manager believes would not be appropriate. Or the Contractor has allowed a particular quantity of labour or a planned output, which the Project Manager believes is not practicable.

- *it does not show the information which this contract requires*
  It may not show Key Dates or access dates, method statements, or any indication of float.

- *it does not represent the Contractor's plans realistically*
  The Project Manager may not believe the Contractor can complete an operation within the time he has allowed, or he could be showing planned Completion, which the Project Manager believes is too early and therefore unrealistic.

- *it does not comply with the Works Information*
  The Contractor may not have allowed for a constraint within the Works Information. Or the programme may simply be in the wrong format. For example, the Works Information may prescribe that the programme has to be compiled and submitted using a particular brand of software, but the Contractor has used a different brand.

It has to be said that in many cases the Project Manager does not accept the programme because the Contractor has misunderstood the contract requirements, has not included all the stated constraints, or has not taken proper account of all the factors included in the Works Information.

It is important that the Project Manager either accepts or does not accept the Contractor's programme and if he does not accept it he clearly states why he does not accept it. There is no default position if the Project Manager fails to approve or reject the Contractor's first programme or subsequent submitted programmes. If he does not accept the programme, and it is not for a reason stated in the contract that is in Clause 31.3, then it is a compensation event, though the Contractor is not prevented from commencing the Works so it is difficult to envisage what the Contractor would be seeking to recover from the breach through the compensation event process.

It has been known for Project Managers to respond to the submission of the programme either by making no comment at all within the two weeks they have to respond, or by stating that the programme is accepted on condition that the Contractor makes certain amendments to it.

In the first example, as the ECC does not provide for deemed acceptance where there is no reply to the Contractor's submission, if the Project Manager does not make a comment within the two weeks, the programme must be assumed as not accepted, and as the Project Manager has not stated a reason for not accepting, he has withheld acceptance for a reason not stated in the contract, so it is a compensation event.

In the second example, if the Project Manager states that the programme is accepted subject to amendments, then essentially he is not accepting it as it still has to be amended before he will accept it. Again, Clause 31.3 will apply in terms of non-acceptance.

In both cases the Project Manager has failed in his obligations under the contract.

Note that acceptance of a programme which shows the Contractor completing late does not mean that the Project Manager accepts the delay; it merely acknowledges that the delay is reasonably reflected within the programme.

3.10 Revised programmes

The timing for submission of revised programmes for acceptance is defined by Clause 32.2:

- *within the period for reply after the Project Manager has instructed him*
  For example, the Project Manager could instruct the Contractor to submit a programme to enable an issue to be discussed in a risk reduction meeting.

- *when the Contractor chooses*
  Again, the Contractor may feel it beneficial to submit a programme to be discussed in a risk reduction meeting.

- *and in any case at intervals no longer than that stated in Contract Data Part 1*
  This is the longest period for submission of a revised programme.

The revised programme shows:

- *the actual progress achieved on each operation and its effect upon timing of the remaining work*
  The progress achieved may be shown on the programme itself, or appended to it in the form of a progress report. Operations which are not yet completed will normally be shown as percentage complete, this being the progress achieved to date against the total work within the operation on the date the assessment is made. It is useful if the Contractor and the Project Manager agree the progress statement before the revised programme comes into being. The submitted revised programme is then an undisputed statement of fact.

  The revised programme should reflect planned v actual progress including early or delayed start, early or delayed Completion of each operation.

  It is important that, if the Contractor states a percentage completed to date, he should show actual progress achieved rather than time expired to date. Also, the Contractor should not just make statements such as 'operation on programme' as this may be misleading. An operation on programme does not necessarily mean that the operation is not delayed, for example the Contractor is expecting to complete an operation late, then stating the words 'operation on programme' can be interpreted as either in accordance with the contract, or progressing late in accordance with his expectations.

It is also important to recognise that many operations may not progress on a 'straight line' basis; for example, building a brick wall may initially involve work and resources in setting out and building corners, with the bulk areas following on.

- *the effects of implemented compensation events*
  The June 2006 amendment to the contract changed the sub-clause from 'the effects of implemented compensation events and of notified early warning matters' to 'the effects of implemented compensation events'. It is perhaps curious that the effects of notified early warning matters are not required to be shown on revised programmes.
  Note that a compensation event is not 'implemented' until the quotation has been accepted or assessed by the Project Manager. Whilst the Completion Date is not changed until a compensation event has been implemented, there is a danger in only showing compensation events that have been implemented as the revised programme is not reflecting reality.

Example

The Project Manager instructs the Contractor not to paint the walls to certain rooms of a new building as he is considering an alternative finish which he will instruct at a later date.

The following week, the Contractor is due to submit a revised programme, but he has only just submitted his quotation to the Project Manager for the omission of the painting, so it is probably a couple of weeks before the Project Manager replies and the compensation event is implemented. Should the Contractor in the meantime be showing the walls as to be painted or not to be painted as the compensation event has not yet been implemented? It should show the walls as not to be painted as it reflects reality.

- *how the Contractor plans to deal with any delays and to correct notified Defects*
  The Contractor should show in his revised programme any delays which may or may not be caused by him, and also time needed to correct his own Defects.

- *any other changes which the Contractor proposes to make to the Accepted Programme*
  An example of this could be that the Contractor resequences part of the work or proposes to use a different method or different Equipment to that specified by him in a previous method statement.

It is therefore very much a 'living' programme. If the Contractor does not submit a programme which the contract requires, then the Project Manager can assess compensation events, without receiving a quotation from the Contractor (Clauses 64.1 and 64.2).

The Accepted Programme:

- effectively provides an agreed record of the progress of the job and where the delays have come from;
- provides a realistic base for future planning by both the Contractor and Project Manager;
- is the base from which changes to the Completion Date are calculated;
- is the base from which additional costs are derived.

Because each operation has a method statement and resources attached to it, the change in resources can be calculated and hence the change in costs. Because information is so much more transparent, there is more scope for working together.

3.11 Record keeping

One of the most important, but often understated, elements of managing a construction contract and its progress and any difficulties, whether one is a Contractor or a Project Manager, is the keeping of concise and timely records. It is beneficial if the records are agreed by the Contractor and the Project Manager. These are then undisputed records of fact. It is then only the interpretation of these facts which might subsequently be at issue.

A good record keeping system will enable the Contractor or the Project Manager to notify early warnings promptly, assist the preparation of programmes, enable payments to be made under Options C to E, and also prove invaluable in pricing and assessing compensation events, all of which have time scales applied to them, so readily available records are a must.

Records can include:

- Site diaries;
- Site meeting minutes;
- photographs including dates;
- labour returns;
- Equipment returns;
- weather records;
- correspondence;
- financial accounts and records.

3.12 Clause 34: Instructions to stop or not to start work

The Project Manager has the authority to give instructions to the Contractor to stop or not to start work for any reason and may later instruct him to re-start it. The instruction would be a compensation event (Clause 60.1(4)). However, if the instruction was required due to a fault of the Contractor, for example unsafe working, the Project Manager should instruct the Contractor that the Prices, the Completion Date and the Key Dates are not to be changed.

3.13 Clause 35: Take-over

Once the Contractor has completed the Works, the Employer takes over within two weeks. This occurs even if the Contractor completes the Works early. However, there is an optional statement within Contract Data Part 1 that 'the Employer is not willing to take over the Works before the Completion Date'. If this is selected then if the Contractor completes the Works early he still has responsibility for the Works until the Completion Date.

Whilst many contracts have express provision for partial possession by the Employer, with the consent of the Contractor, which shall not be unreasonably withheld, the ECC does not, though under Clause 35.2 the Employer may use any part of the Works before Completion has been certified. Take-over of that part is deemed to have occurred unless otherwise stated in the Works Information, or it has been done to suit the Contractor's method of working.

Note that, if the Project Manager certifies take-over of a part of the Works before Completion and the Completion Date, then it is a compensation event under Clause 60.1(15); if the Employer takes over the Works before Completion but after the Completion Date, then the Contractor is already late in completing and therefore that would not be a compensation event. There is no provision for the Contractor to give or refuse consent to the Employer taking over the Works.

The Project Manager certifies the date of take-over, or partial take-over, within one week of it taking place.

3.14 Clause 36: Acceleration

Acceleration in many contracts normally means increasing resources, working faster or working longer hours so that Completion can be achieved by the Completion Date, in effect the Contractor is catching up to recover a delay. However, within the ECC acceleration means bringing increasing resources, working faster or working longer hours so that Completion can be achieved before the Completion Date.

The Project Manager may instruct the Contractor to submit a quotation for acceleration. He also states any changes to the Completion Date to be included within the quotation. The quotation must include proposed changes to the Prices, and a programme showing the earlier Completion Date and Key Date. As with quotations for compensation events, the Contractor must submit details within his quotation, however the contract does not prescribe how the quotation is to be priced, for example based on Schedule of Cost Components.

There is also no remedy if the Contractor does not submit a quotation, or if the Contractor's quotation is not accepted by the Project Manager in that he could make his own assessment. Acceleration can only be undertaken by agreement between the Project Manager and the Contractor and cannot be imposed on the Contractor or any assessment imposed upon him without his agreement.

When the Project Manager has accepted the quotation for acceleration, he changes the Prices, Completion Date and Key Dates and accepts the revised programme.

4 Testing and Defects

4.1 Introduction

Testing and Defects, including carrying out tests, inspection and the correction of Defects are covered within the ECC by Clauses 40.1 to 45.2 with further references within the Main Option clauses.

Testing within this contract is defined through the Works Information and the applicable law.

Any Materials, facilities and samples for testing, including who provides them, are also stated within the Works Information.

4.2 Quality management and control

Introduction

Quality management is a major management function within the construction industry and its various companies.

Unless a construction company can guarantee its Employers a quality product, it cannot compete effectively in the modern construction market.

Quality often stands alongside cost as a major factor in Contractor selection by Employers, as well as determining the efficiency of processes that the contractor adopts for Site operations. To be competitive and to sustain good business prospects, these quality systems need to be efficient and evidential.

The role of quality management for a construction company is not an isolated activity, but intertwined with all the operational and managerial processes of that company.

The modern concept of quality is considered to have evolved through three major stages over many years.

These stages are as follows.

(i) Quality control and inspection

Inspection is the process of checking that what is produced is what is required. It is about the identification and early correction of Defects.

Quality control introduces inspection to stages in the Works ensuring that it is undertaken to specified requirements. Quality control is done on a sampling basis dictated by statistical methods, ie sampling concrete cubes, mock-ups, etc. The quality is measured by comparing the work actually carried out against the sample initially provided.

(ii) Quality assurance

Quality assurance is the process of ensuring that standards are consistently met thereby *preventing* Defects from occurring in the first place.

'Fit for purpose' and 'right first time, every time' are the principles of quality assurance.

(iii) Total quality management

This is based on the philosophy of continuously improving goods or services. The key factor is that everyone in the company should be involved and committed from the top to the bottom of the organisation.

The successful total quality managed company ensures that their goods and services can meet the following criteria:

- be fit for purpose on a consistently reliable basis;
- delight the customer with the service which accompanies the supply of goods;
- supply a quality product that is so much better than the competition that customers want it regardless of cost. It has to be said that whilst the construction industry prides itself on the ability to deliver quality projects, it is virtually unknown for Employers to disregard cost!

The pursuit of total quality is seen as a never-ending journey of continuous improvement.

The earliest and most basic form of quality management is quality control. This is described under the headings of:

1. quality control;
2. controlling quality;
3. quality control implemented in construction.

Quality control

The term 'quality control' is defined by an interpretation of its elements: 'quality' and 'control'.

Quality

The term 'quality' is often used to describe prestige products such as expensive jewellery and motor cars. However, although applicable to these items, the term 'quality' does not necessarily refer to prestigious products but merely to the fitness of the product to the customer's requirements. Quality is therefore described as meeting the requirements of the customer.

Control

The concept of being 'in control', or having something 'under control', is readily understood – we mean we know what we intend to happen, and are confident that we can ensure that it does.

Quality control is primarily concerned with Defect detection and correction. The main quality control technique is that of inspection and statistical quality control techniques (ie sampling) to ensure that the work produced and the Materials used are within the tolerances specified. Some of these limits are left to the inspector's judgement and this can be a source of difficulty.

The major objectives of quality control can be defined as follows:

- to ensure the completed work meets the specification;
- to reduce customers' or Employers' complaints;
- to improve the reliability of products or work produced;
- to increase customers' or Employers' confidence;
- to reduce production costs.

Controlling quality

The central feature to all quality control systems is that of inspection. To be effective the construction process requires that work items to be inspected must be catalogued into a quality schedule.

In the case of construction, inspection takes two forms: that which is quantifiable, eg line, levels, verticality and dimensions; and that which is open to the inspector's interpretation, eg cleanliness, fit, tolerances and visual checks.

There are some precise quantified inspections including the commissioning of Plant and Machinery, pressure tests in pipe-work and strength tests on Materials such as concrete.

Statistical methods

These methods of quality control are based on the need to sample. In many of the processes of manufacture and construction the scale of the operation is too large to have 100 per cent inspection and therefore sampling techniques are employed.

The main techniques in statistical quality control are as follows:

- *Acceptance sampling*, based on probability theory, allows the work to continue if the items sampled are within pre-determined limits.

- *Control charts* that compare the results of the items sampled with the results expected from a 'normal' situation. Usually the results are plotted on control charts which indicate the control limits.

In construction, it is the quality of Materials that is normally controlled by statistical methods, the most common being that of the cube strength of concrete.

Quality control implemented in construction

Traditionally there are two sets of documents that are used to determine the required quality of a construction project. These are the specifications and the contract drawings. The Contractor uses these two documents during the Site operations stage of any project to facilitate 'quality' construction.

The process of actual construction is dissimilar to that of a production line in that there are no fixed physical and time boundaries to each operation of the process, hence the positioning and timing of quality inspections cannot be predetermined. In construction, quality checks are undertaken as each operation or sub-operation is completed.

The majority of quality checks are undertaken visually. Visual quality checks of each section of construction are undertaken by the Contractor's engineers and foremen and then by the resident engineers and inspectors to ensure it complies with the drawings and specification.

Quantifiable quality checks are also made during the construction stage. These include testing the strength of concrete cubes, checking alignment of brickwork and commissioning of services installations. The results of these quality checks are recorded and passed to the appropriate party.

Notwithstanding the existence of quality assurance and the emergence of total quality management, most Employers still engage inspectors to reassure themselves. However, the impact and importance of the Employer's inspectors is much reduced in a quality assured or total quality managed company.

Quality assurance

Whereas quality control focuses on Defect detection and correction during the Works, quality assurance is based on Defect prevention.

Quality assurance concentrates on the production or construction management methods and procedural approaches to ensure that quality is built into the production system.

Quality assurance is described under the following headings:

- evolution of quality assurance from quality control;
- definition of quality terms;
- quality assurance standards;
- developing and implementing a QA system;
- quality assurance in construction.

4.3 The ECC requirements

Whereas most of the interactions within the contract are between the Contractor and the Project Manager, within Section 4 they are between the Contractor and the Supervisor, the Project Manager playing a fairly minor role, only being involved in respect of arranging access for the Contractor to correct a Defect, assessing costs incurred by the Employer in repeating a test or inspection and assessing the cost to the Contractor of correcting a Defect when he is not given access after the Employer has taken over the Works.

The Supervisor, in conjunction with the Contractor, notifies tests as and when required and carries out the various inspections as required by the contract. It is important to recognise, as previously stated in the first chapter, that the Supervisor may be appointed from the Employer's own staff or may be an external consultant, but he represents the Employer in respect of testing and Defects, he is not an assistant or deputy to the Project Manager.

The Contractor and the Supervisor are required to notify the other of tests and inspections before they start, and also the results of those tests. If a test shows that any Works has a Defect, the Defect is corrected and the test repeated. The Project Manager assesses the cost incurred by the Employer in repeating a test or inspection after a Defect is found and the Contractor pays the amount assessed.

Under Options C, D and E, when the Project Manager assesses the cost incurred by the Employer in repeating a test or inspection after a Defect is found, he does not include the Contractor's cost of carrying out the repeat test or inspection.

Note, however, that under Options C, D and E, the Contractor may be paid for correcting a Defect.

The Supervisor also has the right to watch any test being done by the Contractor.

The Supervisor is also obliged to do his tests and inspections without causing unnecessary delay to the work, or to a payment which is conditional upon a successful test. The term 'unnecessary delay' is subjective; clearly it must be expected that there may be some delay, whist the Supervisor does his tests and inspections. If the test or inspection causes unnecessary delay, then it is a compensation event (Clause 60.1(11)).

If a payment is conditional upon a test or inspection being successful it becomes due for payment at the later of the Defects Date and the end of the last Defect Correction Period if:

• the Supervisor has not done the test or inspection; or
• the delay to the test or inspection is not the Contractor's fault.

Should the contract require Plant and Materials to be tested before delivery, for example structural or electrical testing, they are not brought to the Working Areas until the Supervisor has notified the Contractor that they have passed the test or inspection.

Contract:	SUPERVISOR/CONTRACTOR NOTIFICATION
Contract No:	S/CN No..

To: Contractor

Type of Notification/Instruction:	Date	Time
☐ **Notification of test or inspection**	....	....
☐ **Notification of result of test or inspection**	....	....
☐ **Notification of identified Defect**	....	....

Instruction to search (Supervisor only):
Location and Details:

Result of test or inspection:
Test: PASSED/FAILED
Reason for failure:

It is planned to correct the Defect and retest on (date) ....

Defect corrected Yes/No   **Date:........................**
Other comments or action:

Signed: (Supervisor/ Contractor) .... Date: ....

Copied to:

Contractor ☐ Project Manager ☐ Supervisor ☐ File ☐ Other .... ☐

Figure 4.1 Sample Supervisor/Contractor notification

4.4 Defining a Defect

A Defect is defined within the contract as:

* *a part of the Works which is not in accordance with the Works Information*
  The Works Information provides the reference point for what the Contractor has to do to provide the Works. If the work done by the Contractor does

not comply with the Works Information, then unless the Defect is accepted (Clause 44), the Contractor is obliged to correct the Defect. In most cases the work done will fall short of the Works Information, but as the words 'not in accordance with' are used, work which exceeds the Works Information requirement would also be a Defect. Clearly, the Project Manager may wish, having discussed with the Employer, to accept such a Defect!

A Defect may not necessarily mean that the work is not fit for purpose; the Defect could simply be a colour variation, for example the Works Information stated a certain shade of red paint and the Contractor used a different shade of red paint, in which case a Defect exists as that part of the Works was not in accordance with the Works Information. In such a case, unless the shade of red is a particular corporate colour or a stipulation of a regulatory authority, it may be prudent to accept the Defect.

It is also possible that work is 'defective' but as it complies with the Works Information, it is not a Defect! For example, the specification within the Works Information may stipulate use of a particular material which then fails. In that respect the work is defective but as it complies with the Works Information, it is actually not a Defect.

- *a part of the Works designed by the Contractor which is not in accordance with the applicable law or the Contractor's design which the Project Manager has accepted*
  If the Project Manager accepts the Contractor's design under Clause 21.2, then subsequently the Contractor either does not comply with the applicable law, or changes the design, then again that is a Defect. If the Contractor wishes to change his design then he has to re-submit the new design to the Project Manager for his acceptance.

4.5 The quality-related documents within the Works Information

Scope of Works

The Scope of Works must define clearly what is expected of the Contractor in Providing the Works. It must cover all aspects of the work, either specifically or by implication, otherwise it may be deemed to be excluded from the contract.

A comprehensive description should therefore be considered for the Scope of Works, which should be supplemented by specific detailed requirements in the form of drawings and specification. If reliance is placed solely on the Scope of Works, there is a danger that an item may be missed and could be the subject of later contention.

Whilst the Scope of Works must clearly describe the boundaries for the work to be undertaken by the Contractor, it may also describe the Employer's objectives and explain why the work is being undertaken and how it is intended to be used. It says what is to be done and what is to be included in general terms, but not how to do it.

The Scope of Works is the document that a tenderer can look to, to gain a broad understanding of the scale and complexity of the job and be able to judge his capacity as a Contractor to undertake it. It is written specifically for each contract and should be comprehensive, but it should be made clear that it is not intended to include all the detail, which will also be contained in the drawings, specifications and schedules.

Drawings

The drawings should detail all the work to be carried out by the Contractor and be complementary to the Scope of Works and the specification.

Specification

The specification is a written technical description of the character and quality of Materials and workmanship for the work which is to be carried out. It may also refer to the phasing or sequencing of the work to be carried out and also the outcomes to be achieved.

Note that there is no hierarchy or precedence between documents within the Works Information. If there is any ambiguity or inconsistency in or between the documents which form the Works Information, the Project Manager or Contractor notifies the other under Clause 17.1, and the Works Information is changed by the Project Manager to correct the ambiguity or inconsistency. This is then a compensation event (Clause 60.1(1)) which is assessed as if the Prices, Completion Date and Key Dates were for the interpretation most favourable to the party which did not provide the Works Information.

Responsibilities

The responsibilities of the contracting parties and their representatives should be clearly defined in a manner which will leave no doubt as to the obligations that each is accepting. This will also allow the procedures necessary to enable the contract to progress satisfactorily from inception to Completion to be established at the outset of the work.

The contract as a whole will also define the responsibilities for the design, production and programming of the design information, the requirements for design approval, the supply of any free-issue Materials, the issuing of instructions and the form which these instructions are to take, the programming of the Works, the method of measuring and evaluating the work, the circumstances which will constitute a variation to the Scope of Works and the duties of the parties during construction and installation.

Testing take-over and liability for Defects

Various liabilities commence or are released when certain stages have been reached in the work. Such stages are usually acknowledged by the issue of a certificate, which may include certificates of handover and take-over, partial or final Completion, and maintenance period.

Modern-day projects, by their very nature, will require tests to be conducted both during and on Completion of individual installations and in many cases in relation to a number of separate installations which operate in conjunction with each other to meet the design parameters of the project as a whole.

Testing therefore may relate to individual material components, factory testing of completed items of Equipment off Site, pressure testing of individual sections of pipe work, testing of welds or pipe joints, a vast array of electrical and process control tests and tests of environmental conditions or quality of finished product after processing. Testing may also be required for insurance purposes.

Testing needs to be interpreted separately from pre-commissioning or commissioning, as does performance testing, where the whole or part of an installation requires certain preset specified design parameters to be met. The Employer may arrange for commissioning of completed installations or groups of installations to be carried out independently or jointly with the Contractor(s).

While the specification will set out the procedures and parameters for testing and commissioning, the contract will set out the obligations as to how these parameters are to be met and the consequences of failure to meet them. The responsibility for the provision of fuel for testing and testing Equipment also needs to be stipulated.

Performance testing is usually applicable where the design is the responsibility of the Contractor and can include the running of the completed Plant and the checking of product quality and quantity, feedstock consumed, use of energy, waste and by-products, environmental conditions and other aspects such as may be required. In such instances time limits for the rectification of performance Defects must be stated together with penalties for failure.

The documentation required in the form of test certificates, warranties and guarantees to enable proper records to be maintained should be stipulated, as should the effect in relation to the guarantee or warranty period and the extent to which it will apply.

In many cases the awarding of a certificate following testing will establish the date of take-over of the installation or, where applicable, that part ready for commissioning.

Liability for Defects should be defined in the contract together with the period of time for which such liability is to apply and the length of time within which Defects are to be remedied.

Limitation periods

Whilst the ECC has a period from Completion to the Defects Date during which the Contractor is liable for correcting Defects, most countries also have a statutory limitation period for latent Defects which come to light at a later date, beyond which a claim cannot be made for defective work.

In the UK, for example, the time limit for actions founded on simple contract is 6 years and for a specialty contract or deed, the time limit is 12 years, in both cases this is measured from the date on which the cause of action accrued, normally this is taken as the date of Completion.

4.6 Defining testing within the Works Information

Testing can be related to individual Plant and Materials within the Working Areas, factory testing outside the Working Areas, testing of individual sections of work, welds and joints once they have been assembled or a multitude of mechanical and electrical tests.

Whilst the Works Information will set out the procedures and parameters for testing and commissioning, the contract will set out the parties' obligations as to how criteria are to be met and the consequences of failure to meet them.

Performance testing is usually applicable where the design of Plant and Materials is the responsibility of the Contractor. This can include the operation of the completed Plant and the checking of product quality and quantity, fuel consumed, use of energy, waste and by-products, environmental conditions and other aspects such as may be required. In such instances time limits for the rectification of performance Defects must be stated together with penalties for failure.

The documentation required in the form of test certificates, warranties and guarantees to enable proper records to be maintained should be stipulated, as will the effect in relation to the guarantee or warranty period and the extent to which it will apply.

In many cases the awarding of a handover certificate following testing will establish the date of take-over of the installation or, where applicable, that part ready for commissioning.

Liability for Defects should be defined in the contract together with the period of time for which such liability is to apply and the length of time within which Defects are to be remedied.

It is imperative that the Works Information, produced at tender stage and referred to in Contract Data Part 1, defines the full extent and timing of tests and inspections required and the subsequent correction of any Defects.

This is very often not covered in sufficient detail within the Works Information, so it is important for the compiler of the Works Information to consider a number of key questions:

(i) What is the purpose of the test?

What is it intended to prove or disprove? Is the test required before a following piece of work can be carried out by the Contractor or another party, or is it required before the Employer can take over the Works or part of the Works?

(ii) What is to be tested?

What elements or components are required to be tested? Is it to be tested in an assembled or unassembled state?

(iii) How is the test to be carried out, and who is to provide the Materials, facilities and Equipment to do the test?

Whether the test is a well-known 'industry standard' test or not, the exact nature of the test must be clearly defined. Who supplies the Equipment, who pays for the power or fuel for the Equipment, etc? Is a component, for example a concrete cube, to be tested to destruction? It is also important to detail who is to provide the Materials and facilities to do the test and who meets the costs.

(iv) Who should do the test? Who should be present and who should be invited should they wish to attend when the test is taking place?

There may be a requirement for an independent party to carry out the test and provide results.

(v) When and where will the test take place? On Site or off Site?

Clearly state at what stage or on what date the test must take place. Sometimes, tests may have to be carried out on a periodic basis, for example on concrete samples on delivery and as the concrete cures.

(vi) What is the expected result or outcome from the test?

What are we expecting the test to prove? What results do we need to gather once the test has been carried out?

(vii) Who should be advised of the outcome of the test?

The Supervisor will normally be advised of the outcome of a test, but sometimes a Subcontractor or a third party outside the contract may need to be informed.

(viii) What should be done in the event of a successful test, eg certification?

Does a certificate have to be issued following successful Completion of a test? Does that successful outcome mean that the Contractor can proceed to the next stage of the work? Is payment to the Contractor dependent on successful Completion of the test?

(ix) What should be done in the event of an unsuccessful test, eg rectification or replacement, and further testing?

Does the Contractor have to repeat the test? Is the retest done immediately after the first test, or should there be a period before the next test? Is it possible that even though something has failed a test, it may still be considered acceptable, possibly by the Contractor offering a cost saving?

**(x) What about additional testing, ie to confirm a suspected Defect
or the extent of a suspected Defect?**

This may be referred to as a 'search' where the Contractor may be instructed to
'open up' work which is believed to be defective, possibly including the requirement for some form of additional testing.

There may be a requirement for Plant and Materials to be tested off Site before
delivery; this is covered by Clause 41.1.

The main documents within the Works Information in respect of testing and
inspection will normally be the drawings and specification.

4.7 Defaults and remedies

Where the Contractor has failed to comply with the terms of the contract, the
Employer may have the right to take appropriate action to remedy the default
including, where applicable, having the work completed by another Contractor
at the expense of the original Contractor.

The manner of serving notice of default and the financial consequences of the
default should be stated.

Where either party to the contract has defaulted to the extent that financial
redress or litigation is invoked, contemporary records will be required by way of
substantiation. Proper record-taking procedures should already have been established but, as soon as potential default situations become evident, these procedures may need to be reviewed and additional records kept.

4.8 Searching for Defects

Under Clause 42.1, the Supervisor (not the Project Manager, unless the Supervisor has delegated the authority to him) has the authority to instruct the Contractor to search. This action is normally defined as 'uncovering' or 'opening up' in
other contracts. It is usually required in order to investigate whether a Defect
exists, and possibly the cause, and the correcting measures required.

Searching can include uncovering, dismantling, reassembly, providing Materials and samples and additional tests which the Works Information did not originally require.

If the Supervisor instructs the Contractor to search for a Defect and no Defect
is found, this is a compensation event (Clause 60.1(10)). However, if the search
is needed only because the Contractor gave insufficient notice of doing work
obstructing a required test or inspection, then it is not a compensation event.

Until the Defects Date, the Supervisor is obliged to notify the Contractor of
every Defect as soon as he finds it, and the Contractor is obliged to notify the Supervisor as soon as he finds it. Whilst the requirement for the Supervisor to notify the
Contractor of each Defect is useful as the Contractor may not have been aware of
the Defect or that it was a Defect, so he can correct it immediately, the obligation

for the Contractor to notify the Supervisor of each Defect as soon as he finds it is somewhat cumbersome, as the Supervisor probably does not want or need to know about every Defect, particularly if the Contractor is already correcting it.

4.9 Correcting Defects

First, it is important to stress that the Contractor is responsible for meeting the quality standards required by the contract and has an obligation to correct Defects whether or not the Supervisor has notified him of them. In that respect he should adopt a quality assurance/quality control system which prevents Defects occurring and corrects them promptly if they do occur.

The ECC has two periods for correction of Defects which are identified within Contract Data Part1:

1. The period between Completion and the Defects Date for Defects which were not apparent or were not notified at Completion, or were notified, but did not prevent the Employer from using the Works. This period is set within Contract Data Part 1, and is normally 52 weeks. This is similar to the 'Defects Liability Period', 'Rectification Period' or 'Defects Notification Period' in other contracts, the period within which the Contractor is initially liable for correcting Defects.

2. The 'Defect Correction Period' is the period within which the Contractor must correct a notified Defect, failing which the Project Manager assesses the cost incurred to the Employer of having the Defects corrected by other people and the Contractor pays this amount. Note that, in this respect, although the Contractor has already paid the amount, the Employer may wish to leave the Defect uncorrected. The Defect Correction Period starts at Completion for Defects notified before Completion and when the Defect is notified for other Defects. The Defect Correction Period is not effective prior to

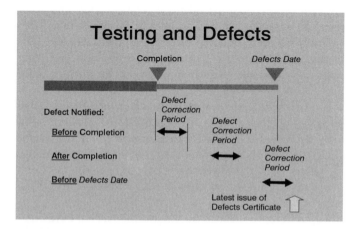

Figure 4.2 Defect Correction Period illustration

Contract:	**DEFECTS NOTIFICATION**
Contract No:	D/N No...

The Contractor is notified of the following Defects:

Description	Date Corrected

The Defect Correction Period is:

Certified all Defects
corrected................................(Supervisor)   Date: ....

Copied to:

Contractor ☐   Project Manager ☐   Supervisor ☐   File ☐   Other .... ☐

Figure 4.3 Sample Defects notification

Completion; therefore if a Defect becomes apparent prior to this, the Contractor has an obligation to correct it in readiness for Completion.

Some Employers may wish to designate different Defect Correction Periods for different Defects depending on their urgency; Contract Data Part 1 provides for this. A typical example of a Defect Correction Period for a water company upgrading domestic water supplies is:

- Type A Defects – 4 hours

  - major leaks
  - Defects related to health and safety.

- Type B Defects – 24 hours

  - minor leaks.

- Type C Defects – 7 days

  - cosmetic Defects.
  - Defects which are not Type A or B.

Defects which may prevent the Employer from using the work must be corrected before a Completion Certificate can be issued (Clause 11.2(2)).

If a Defect is required to be corrected after the Employer has taken over the Works, the Project Manager arranges for the Employer to allow access to the Contractor. The Defect Correction Period in respect of the Defect does not commence until the necessary access has been provided.

4.10 The Defects Certificate

The Defects Certificate is issued to the Project Manager and the Contractor by the Supervisor at the later of the Defects Date and the end of the last Defect Correction Period. Some practitioners suggest that as the Project Manager issues the Completion Certificate, he should also issue the Defects Certificate, which is common with most other contracts, but by requiring the Supervisor to issue the Defects Certificate, the ECC reinforces the role and responsibility of the Supervisor in dealing with Defects on behalf of the Employer.

Whilst many contracts require any Defects to be corrected before the Defects Certificate, or its equivalent is issued, under the ECC the Defects Certificate is issued by the Supervisor at the appropriate time and may either list Defects which the Contractor has not corrected, or a statement that there are no outstanding Defects.

Note that the Defects Certificate is not conclusive in that if a Defects Certificate states that there are no Defects, it does not prevent the Employer from changing the Employer's rights should a Defect arise later or if the Supervisor did not find or notify it. However, the Employer may wish to take action against a Supervisor who did not carry out the level of inspection he had paid him to do!

The Supervisor will normally only list what are usually defined as 'patent Defects', ie those which are observable from reasonable inspection at the time, examples being a defective concrete finish or an incorrect paint colour, and may not include what are usually defined as 'latent Defects', which may be hidden from reasonable inspection and may come to light at a later date, examples being some structural Defects.

The Contractor's and others' liability for correction of latent Defects and other costs associated with them will be dependent on the applicable law, and liability may remain despite the issue of the Defects Certificate.

4.11 Accepting Defects

A Defect is defined as 'a part of the Works that is not in accordance with the Works Information or a part of the Works designed by the Contractor which is not in accordance with the applicable law or the Contractor's design which the Project Manager has accepted' (Clause 11.2(5)).

In the event that a Defect becomes apparent, there are two choices:

1.   The Contractor corrects the Defect so that the part of the Works is in accordance with the Works Information or the design is in accordance with the applicable law or the Contractor's design.
2.   The Contractor and Project Manager may each propose to the other that the Works Information should be changed so that a Defect does not have to be corrected. Note that 'each proposes to the other' requires the acceptance of the Contractor and the Project Manager, the latter probably discussing the matter with the Employer.

Example

The Contractor has carried out a large area of wall tiling in a proposed new railway station. Whilst visually the quality of the work appears to be excellent, the tiles have not been laid to the required tolerance within the Works Information and therefore the work is defective. The railway station is due to be completed in two weeks, so if the Contractor has to take down the wall tiles, re-order another batch of tiles and then lay the new tiles to the required tolerance, that could take a considerable time and possibly prevent completion of the Works or that specific part of the Works.

In this case, provided both the Contractor and the Project Manager are prepared to consider changing the Works Information, then the Contractor provides a quotation to the Project Manager stating a financial saving (reduced Prices), based on him not having to correct the Defect or an earlier Completion Date or both, and if the Project Manager accepts the quotation then the Works Information is changed and the Prices and/or Completion Date are changed in accordance with the quotation.

This is obviously not an option where the design is not in accordance with the applicable law.

If the Contractor and Project Manager consider this change, the Contractor submits a quotation for reduced Prices or an earlier Completion Date, or both. If the Project Manager accepts the quotation he gives an instruction to change the Works Information, the Prices and the Completion Date.

4.12 Uncorrected Defects

The Project Manager arranges for the Employer to allow access to parts of the Works taken over in order to correct a Defect. As stated above, if the Contractor does not correct the Defect within the Defect Correction Period, the Employer assesses the cost of having the Defect corrected by other people and the Contractor pays this amount.

However, if the Contractor is not given access to correct a Defect, the Project Manager assesses the cost to the Contractor of correcting a Defect and the Contractor pays this amount. This is an unusual clause in that most contracts provide for the Contractor to have the opportunity, as well as the obligation, to correct his own Defects; if he does not take the opportunity, he is liable for the cost of someone else correcting it. In this case, the Contractor could be denied the right to re-enter the affected area and be subject to the Project Manager's subjective opinion as to how much cost the Contractor would have incurred in correcting the Defect(s) if he had been allowed access.

4.13 Liability for the cost of correcting Defects

The Contractor is liable for correction of Defects; however, note that under Options C, D, and E where the Contractor is reimbursed his Defined Costs, Disallowed Cost is only for the cost of:

* correcting Defects after Completion;
* correcting Defects caused by the Contractor not complying with a constraint on how he is to Provide the Works stated in the Works Information,

From the first bullet point, the Employer pays for Defects to be corrected before Completion but not after Completion. There is no limit to the type or size of Defects, or why the Defect occurred, eg through the carelessness of the Contractor.

Whilst some may see this as the Contractor's right to repeatedly attempt to get the work right at the Employer's expense, it must be remembered that in the case of Options C and D these are target contracts, therefore if the Contractor is paid for correcting a Defect, not only is this bad for his reputation, but as it is not a compensation event he is not being given additional time to correct the Defect and the additional cost paid will reduce his potential entitlement to Contractor's share at Completion.

From the second bullet point, if the Works Information prescribes a particular method of Providing the Works and the Contractor does not comply, any associated costs would be disallowed.

5 Payment

5.1 Introduction

Payment, including assessment, certification and payment of amounts due is covered by Clauses 50.1 to 52.1, with further references within the Main Option clauses.

There are separate payment provisions within the ECC, dependent on the Main Option chosen, each option referring to two main terms:

(i) The Prices

These are the various elements that make up the total Price for carrying out the work and will be in the form of an activity schedule, bills of quantities, target or some other document dependent on the Main Option selected.

(ii) The Price for Work Done to Date

This term is used in making the assessment of amounts due to the Contractor, and is again dependent on the Main Option chosen:

- Option A: Total of the Prices for completed activities;
- Option B: Quantity of work multiplied by the bill rate, proportion of lump sums;
- Option C: Defined Cost forecast to have been paid by the next assessment date plus Fee;
- Option D: Defined Cost forecast to have been paid by the next assessment date plus Fee;
- Option E: Defined Cost forecast to have been paid by the next assessment date plus Fee;
- Option F: Defined Cost forecast to have been paid by the next assessment date plus Fee.

5.2 Option A: Priced contract with activity schedule

The Prices are the lump sums for each of the activities on the activity schedule unless later changed in accordance with the contract.

The Price for Work Done to Date (PWDD) is the Total of the Prices for completed activities, and each completed activity which is not in a group (minor activities may sometimes be grouped together).

It is important to note that the activity schedule is the basis for the Contractor to be paid, it is not Works Information or a schedule of what the Contractor is intending to do in Providing the Works.

Example

The Contractor has been awarded an Option A contract for the construction of 10 houses in accordance with a design produced by the Employer. However, within the activity schedule the Contractor submitted with his tender, he has not specifically included an activity for the roof covering to the houses, though the roof covering is clearly detailed within the Works Information.

The roof covering is deemed to be included within the Contractor's Price and therefore must be assumed to have been included within the activity schedule. The same rule would apply to an Option C contract.

The activity schedule should also relate to operations on the Contractor's programme, and if possibly due to a change of method of working it does not, the Contractor must submit a revision to the activity schedule to the Project Manager for acceptance.

There will usually be more activities on the activity schedule than operations on the programme. This is because, for an operation which spans across a number of months, the Contractor will include a number of activities.

Example

On the Contractor's programme, he has a drainage operation to the West elevation of the Site, which will be carried out during the months of February, March and April. In order to optimise his use of resources, he will excavate the drain trench in February, lay the pipe in the trench in March, and backfill the trench in April.

His activity schedule could then look like this:

Drainage to West Elevation of Site
Excavate trench £12,200.00
Lay pipe £8,700.00
Backfill trench £6,600.00

It is vitally important that when the Contractor submits the activity schedule each activity is clearly defined to prevent debates between himself and the Project Manager about what an activity is and whether it has been completed.

The Price for Work Done to Date is the Total of the Prices for each completed activity in the activity schedule. NB The contract does not provide for incomplete activities to be paid; it is 'all or nothing'.

A completed activity is one without Defects which would either delay or be covered by immediately following work. So, for example, if the activity was to install a 30 metre length of pipework on and including supports and the pipework and the supports were all installed but there was a Defect in one of the supports, then provided it does not delay or is covered by immediately following work, it is a completed activity, but with a Defect that needs to be corrected. If some of the supports referred in the activity schedule are not yet installed, one should assess this as an activity which has not been completed.

The Contractor is not entitled to be paid for activities which are not complete at the assessment date. Assessing the Contractor's right to payment under the contract is therefore different to the traditional approach of valuing the Works carried out to date.

If the effect of a compensation event is to delay the Completion of an activity, the Contractor is not entitled to be paid a proportion of the Price for that activity, the cost of the delayed payment having to be priced within the quotation for the compensation event that caused the delay.

Activities may include setting up and clearing the Site, time-based preliminaries and activities such as design as well as physical activities on Site. It is important that the activities on the activity schedule are clearly defined not only in terms of what the work is, but where it is, so that their Completion is not in doubt (see activity schedule example below).

The traditional method of forecasting cash flow is by considering the programme and pricing month by month against the programme, inserting values against the events as they occur, allowing for delivery of Materials as they are expected to arrive, and this is essentially the way that Option A works.

In practice, however, particularly for a more complex element such as brickwork, whilst the labour output may be fairly constant, the Contractor may have to allow for all the material being delivered to Site prior to starting work, therefore the activity schedule may have to be 'front loaded' to allow for these early deliveries, however it is important to differentiate this from front loading to create an overpayment to the Contractor.

Payment for large items

The same could apply for other trades such as structural steelwork and mechanical and electrical installations, where large quantities of material are delivered to Site for fixing at a later date.

Example

An activity schedule which incorporates a steel frame valued at £121,000 could be set out as follows:

Steel frame at Subcontractor's premises awaiting delivery	£80,000
Delivery of steel frame to Site	£10,000

Erection of steel to:

GL 1 – 4	£8,000
GL 4 – 8	£8,000
GL 8 – 12	£11,000
Paint steel frame	£4,000

It is also important on projects with preliminaries, to correctly allocate these items to the activity schedule, as they could contribute as much as 10 to 15 per cent of the Total of the Prices.

Payment for preliminaries

Preliminaries would normally comprise the following:

(i) Fixed preliminaries

These would comprise preliminaries which may be single items, not directly influenced by progress on Site.

Typical examples are delivery of temporary accommodation and other Equipment to Site, connection of telephones and other temporary services.

These should be identified within the activity schedule as single activities, eg 'delivery of site accommodation' or a group of activities identified as 'set up Site'. When the activities are completed, they are included within the next assessment.

(ii) Time-related preliminaries

These comprise preliminaries which are based on time on Site rather than having any direct relationship to the quantity of work carried out.

Typical examples are Site management salaries, Site accommodation hire charges, and scaffolding.

These should be identified within the activity schedule as time-based activities, eg 'one month hire of temporary accommodation'. Again, when the activities are completed, they are included within the next assessment.

(iii) Value-related preliminaries

These would consist of preliminaries which are directly related to how much work is being carried out at any particular time.

Typical examples are mechanical Equipment, eg excavators, compressors, mixers.

These, which are not always referred to as preliminaries, should be included within the activities on which they are being used rather than as stand-alone activities.

Design-related activities

On Option A contracts where the Works Information states that the Contractor is to design all or parts of the work, he should consider carefully how and when he requires payment for this activity.

The Contractor should not just include an activity titled 'Design' as there are many aspects of design, and one could argue that design is not completed until the project is completed, so he could deny himself payment for a considerable period.

Ideally, the design-related activities should be linked to various deliverables rather than generic activity descriptions such as 'prepare drawings'.

The various elements of design could be related to, for example, the various stages within the RIBA Plan of Work, or in its simplest form there could be separate activities for the design of various elements of the project, and for the granting of planning approval, etc.

5.3 Option B: Priced contract with bill of quantities

The Prices are the lump sums and the amounts obtained by multiplying the rates by the quantities for the items in the bill of quantities.

The Price for Work Done to Date is the total of:

- the quantity of work which the Contractor has completed for each item in the bill of quantities multiplied by the rate; and
- a proportion of each lump sum which is the proportion of the work covered by the item which the Contractor has completed.

As the quantity of work done is the actual quantity completed, Option B is a remeasurement contract. However, where a change in quantity is related to a compensation event, that change is dealt with separately and the resulting amount fed back into the bill of quantities as a lump sum, a changed rate, or a changed quantity as appropriate.

The bill of quantities is a document for tendering and payment purposes; it is not Works Information or Site Information, so it does not prescribe what the Contractor has to do. It is prepared by the Employer, and the Contractor is assumed to have taken the bill of quantities as correct, and a true reflection of the work to be carried out; therefore, unlike Option A, if an item is missing from the bill of quantities, then it is a compensation event.

Example

The Contractor has been awarded an Option B contract for the construction of 10 houses in accordance with a design produced by the Employer. However, within the bill of quantities there is no item for the roof covering to the houses, though the roof covering is clearly detailed within the Works Information.

Assuming the method of measurement states that roof covering is to be measured as a specific item, rather than included in another item, this would be a compensation event under Clause 60.6 in that there is a mistake in the bill of quantities.

5.4 Option C: Target contract with activity schedule

As with Option A, the Prices are the lump sums for each of the activities on the activity schedule unless later changed in accordance with the contract.

The Price for Work Done to Date is the Defined Cost which the Project Manager forecasts will have been paid by the Contractor before the next assessment date plus the Fee, so the assessment is a combination of Defined Cost paid and yet to be paid by the Contractor. For example, if the assessment date is 1 June, then costs which are forecast to be paid by 1 July are also included.

This is a change to previous editions of the contract which provided for the Contractor only to be paid cost (in NEC2, referred to as 'Actual Cost') that he had paid at the assessment date, the change being introduced to assist the Contractor's cash flow. Whilst the forecast may appear to be no more than a subjective guess, the Contractor has an up-to-date programme and also has probably received many of the invoices which he will be paying before the next assessment date. In practice, however, many Employers particularly in the public sector are resistant to forecasting costs a month ahead and therefore tend to amend the contract through a Z clause to retain the previous wording.

In assessing the amount due if the Contractor has paid costs in a currency different to the currency of the contract, the Contractor is paid Defined Cost in the same currency as the payments made by him. All payments are converted to the currency of the contract applying the exchange rates, in order to calculate the Fee.

Many practitioners question the role of the activity schedule under Option C, when unlike Option A, it is not used for assessing payments. The answer is that the activity schedule under Option C assists the Employer in assessing tenders as they can see how the Contractor has priced his tender.

Again, as with Option A, the activity schedule should relate to operations on the Contractor's programme, and if due to a change of method of working it does not, the Contractor must submit a revision to the activity schedule to the Project Manager for acceptance.

Accounts and records

The Contractor is required to keep the following accounts and records (Clause 52.2) to calculate Defined Cost:

- accounts of payments of Defined Cost;
- proof that the payments have been made;
- communications about and assessments of compensation events for Subcontractors; and
- other records as stated in the Works Information.

The level of checking of the Contractor's accounts and records is at the Project Manager's discretion, some wish to examine each individual cost and its relevant back-up document, some wish only to select certain costs at random, others wishing to only carry out a cursory check of costs. Some Project Managers say that if the parties really are acting in a spirit of mutual trust and co-operation as the contract requires, then there should be no need for a detailed examination of the Contractor's accounts, though this probably needs a quantum leap of faith for many Project Managers and Employers!

It is important to establish the format on which accounts and records are to be presented by the Contractor to the Project Manager, particularly in respect of assessing payments, and this can be detailed in the Works Information, the following alternative methods normally being employed:

- The Contractor submits a copy of all of his accounts and records to the Project Manager on a regular basis.

or

- The Contractor gives the Project Manager access to his accounts and records and the Project Manager takes copies of all the records he requires. Whilst not expressly stated within the contract, the Project Manager should not have to travel too far to inspect them. It has been known for accounts and records to be 'available for inspection' in a different country to the Site, the Contractor stating that they were available for inspection at any time during normal working hours! Again, the location and availability of accounts and records could be detailed within the Works Information.

or

- The Contractor gives the Project Manager password-protected direct access to his computerised accounts and records; the Project Manager can then take copies of any records he requires.

The Contractor's share

The Employer enters the Contractor's share percentage for each share range into Contract Data Part 1, the principle being that the Contractor receives a share of any saving and pays a share of any excess when Defined Cost plus Fee is compared with the target at Completion of the whole of the Works and in the final payment following the issue of the Defects Certificate.

At Completion of the Works, the Project Manager makes a preliminary assessment of the Contractor's share by applying the Contractor's share percentage to the difference between the forecast final Price for Work Done to Date and the forecast final Total of the Prices. A final assessment is made using the final Price for Work Done to Date and the final Total of the Prices.

Whilst NEC3 practitioners often refer to the terms 'pain' and 'gain' in Options C and D, these terms do not actually exist within the contract. The Contractor pays or is paid the Contractor's share.

Example

At Completion of the Works:

The Contractor's share percentages and the share ranges have been entered into Contract Data Part 1 as follows:

share range	Contractor's share percentage
less than 80%	50%
from 80% to 90%	40%
from 90% to 100%	30%
greater than 100%	50%

The Total of the Prices = £3,200,000
The Price for Work Done to Date = £2,400,350

Difference = £799,650

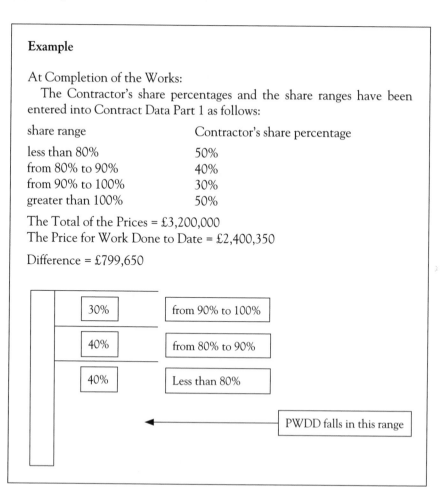

The share range has been set in increments of 10%
10% of £3,200,200 = £320,020

'From 90% to 100%'
£320,000 × 30% = £96,000

'From 80% to 90%'
£320,000 × 40% = £128,000

'Less than 80%'
£159,650 × 50% = £79,825

5.5 Option D: Target contract with bill of quantities

The Total of the Prices is the total of:

- the quantity of the work which the Contractor has completed for each item in the bill of quantities multiplied by the rate; and
- a proportion of each lump sum which is the proportion of the work covered by the item which the Contractor has completed.

Therefore, the target is based on a remeasurement of the work in accordance with the bill of quantities including compensation events.

Completed work is work without Defects which would either delay or be covered by immediately following work.

As with Option C, the Price for Work Done to Date is the Defined Cost which the Project Manager forecasts will have been paid by the Contractor before the next assessment date plus the Fee, and again the Contractor is required to keep accounts and records (Clause 52.2) to calculate Defined Cost.

5.6 Option E: Cost reimbursable contract

The Prices are the Defined Cost plus the Fee.

As with Options C and D, the PWDD is the Defined Cost which the Project Manager forecasts will have been paid by the Contractor before the next assessment date plus the Fee.

5.7 Option F: Management contract

The Prices are the Defined Cost plus the Fee.

As Option E, in terms of the PWDD, except that the definition of Defined Cost is different from Options A, B, C, D and E and is simply the payments due to Subcontractors for work which is subcontracted, plus the costs of any work carried out by the Contractor.

The PWDD is the Defined Cost which the Project Manager forecasts will have been paid by the Contractor before the next assessment date plus the Fee.

5.8 Unfixed Materials on or off Site

A common area of confusion is the payment to the Contractor for unfixed Materials within or outside the Working Areas, ie Materials on or off Site. Unlike other forms of contract, the ECC has no express provisions for payment for such Materials.

In order to be paid for these Materials the Contractor should consider the payment rules for each of the Main Options.

(i) Option A

As the PWDD is based on completed activities, the Contractor should create an activity in the activity schedule for unfixed Materials.

For example, for a structural steel frame, the Contractor could include four activities:

1.   Delivery of steel to Site.
2.   Erection of steel to Gridline 1–10.
3.   Erection of steel to Gridline 10–20.
4.   Paint steel.

(ii) Option B

As the PWDD is based on bills of quantities, appropriate items may be included as method-related charges.

(iii) Options C, D, E and F

As the PWDD is the Defined Cost which the Project Manager forecasts will have been paid by the Contractor before the next assessment date plus the Fee, the Contractor must provide accounts and records to show that he has paid for the Materials or will have paid for them by the next assessment date.

Note that, in respect of unfixed Materials on Site, whatever title the Contractor has to Plant and Materials passes to the Employer if it has been brought within the Working Areas and passes back to the Contractor if it is removed from the Working Areas with the Project Manager's permission.

In respect of unfixed Materials off Site, whatever title the Contractor has to Plant and Materials passes to the Employer if the Supervisor has marked it as for this contract. Also, in respect of marking, the contract must have identified them for payment, and the Contractor must have prepared them for marking as required by the Works Information. This could include setting them aside from other stock, protection, insurance and any vesting requirements.

5.9 Clause 50: Assessing the amount due

The first assessment date is decided by the Project Manager to suit the parties, and will normally be based on the time that the Contractor has been on Site, the Employer's procedures and timing for processing and issuing payments, and the Contractor's payment requirements and internal accounting system.

Clearly, in this respect, some discussion needs to take place between the Project Manager, the Employer and the Contractor in order that a convenient assessment date can be set.

The first assessment must be made within the 'assessment interval' after the starting date, this is normally inserted in the Contract Data as 'four weeks' or 'one calendar month'.

Later assessment dates occur:

- at the end of each assessment interval until four weeks after the Supervisor issues the Defects Certificate;
- at Completion of the whole of the Works.

There is no provision within the contract for a minimum certificate amount.

The Project Manager assesses the amount due at each assessment date, calculating the PWDD using the rules of the specific Main Option.

The amount due to the Contractor is:

- the PWDD;
- plus other amounts to be paid to the Contractor (eg Contactor's share, bonus for early Completion, value added tax);
- less amounts to be paid by or retained from the Contractor (eg retention, delay damages).

In making his assessment, the Project Manager considers 'any application for payment the Contractor has submitted'. Note that in using the phrase 'any application', the Contractor does not have to submit an application. If the Contractor submits an application, the Project Manager considers that; if the Contractor does not, the Project Manager must still make the assessment based on the requirements of the specific Main Option chosen.

It is very unlikely that the Contractor will not submit an application for payment, but this clause does cause some concern among Employers and Project Managers as if the Contractor does not submit an application for payment and the Project Manager makes a mistake in making the assessment and has to correct it, interest on the correcting amount is paid. NB If the correcting amount is an addition, the Employer pays the interest; if a deduction, the Contractor pays the interest.

It is therefore not uncommon for Employers to include within the contract the following Z clause which places the onus on the Contractor to make an application for payment, not on the Project Manager to make the assessment whether or not the Contractor has done so:

> The Contractor submits an application for payment one week before each assessment date. In assessing the amount due, the Project Manager considers the application for payment the Contractor submits. The Project Manager gives the Contractor details of how the amount due has been assessed.

Further assessment dates occur at the end of each assessment interval until four weeks after the Supervisor issues the Defects Certificate, and at Completion of the whole of the Works.

It is essential that the Contractor either submits a first programme with his tender or within the time scale specified within the contract. Failure to do so will entitle the Project Manager to retain one-quarter of the PWDD in his assessment of the amount due. Note that the amount is only withheld if the Contractor has not submitted a programme which shows the information which the contract requires, eg method statement, provisions for float, etc. If the Contractor has submitted a programme which contains all the information that the contract requires, but the Project Manager disagrees with, for example, part of the method statement or the programme has not yet been accepted, then the provision does not apply.

5.10 Clause 51: Payment

The Project Manager is required to certify payments within one week of each assessment date.

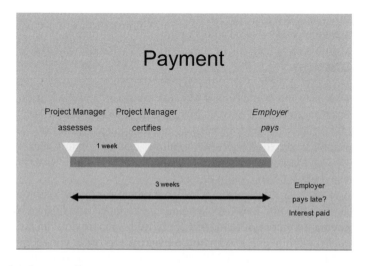

Figure 5.1 Payment illustration

The Employer pays each certified amount within three weeks of each assessment date or, if a different period is stated in the Contract Data, within the period stated. If the certificate is less than the previous one, then the Contractor pays the Employer the amount due. Payments are in the currency of the contract unless otherwise stated.

Interest is paid if a payment is not made or the Project Manager does not issue a certificate which he should have issued. The interest rate is stated in Part 1 of the Contract Data and is assessed on a daily basis from the date the payment should have been made until the date when the late payment is made, calculated using the interest rate in Contract Data Part 1 compounded annually.

The interest due can be calculated on the basis of the following formula:

$$\frac{\text{Payment due} \times \text{interest rate} \times \text{the number of days late}}{365}$$

So if one assumes the following:

- payment due = £120,000
- payment seven days late
- interest rate in Contract Data Part 1 = 3%

The calculation is:

$$\frac{£120,000 \times 3\% \times 7 \text{ days}}{365}$$

Interest due = £69.04

Similarly, interest paid on a correcting amount due applies to later corrections to certified amounts by the Project Manager and on interest due to a compensation event or as determined by the Adjudicator or the tribunal. The date from which interest should run is the date on which the additional payment would have been certified if there had been no dispute or mistake.

5.11 Clause 52: Defined Cost

Contractor's costs which are not included in the definition of Defined Cost are deemed to be included within the Fee percentage.

Note that Defined Cost is at open market or competitively tendered Prices with deductions for all discounts, rebates and taxes which can be recovered, so the Contractor must be able to demonstrate that he has sought alternative Prices. Also, some practitioners believe that this requirement excludes some discounts, eg 2½ per cent cash discount, however the clause states 'all discounts' so there are no exceptions.

The definition of Defined Cost varies according to each Main Option:

- Options A and B – Clause 11.2(22) Defined Cost is the cost of the components in the Shorter Schedule of Cost Components, whether work is subcontracted or not, excluding the cost of preparing quotations for compensation events.
- Options C, D and E – Clause 11.2(23) Defined Cost is the amount of payments due to Subcontractors for work which is subcontracted and the cost of the components in the Schedule of Cost Components, for work which is not subcontracted, less any Disallowed Cost.
- Option F – Clause 11.2(24) Defined Cost is the amount of payments due to Subcontractors for work which the Contractor is subcontracted, plus the cost of any work carried out by the Contractor himself, less any Disallowed Cost.

  The Schedules of Cost Components are specifically excluded in Option F and do not apply.

5.12 Working Areas

The Working Areas are the areas which are necessary for Providing the Works and used only for work on the particular contract.

The Working Areas are the Site, and any other area identified by the Contractor in Contract Data Part 1. An example would be where the footprint of a building filled the Site and therefore the Contractor chooses to use an adjacent piece of land for his Site compound and storage of Materials. In this case, he would identify the additional piece of land as an addition to the Working Areas. The adding of an area may alternatively be submitted as a proposal to the Project Manager during the carrying out of the Works. It is essential that the additional area is necessary for Providing the Works and no other reason.

The significance of the Working Areas is that Defined Cost is identified in the Schedule of Cost Components and Shorter Schedule of Cost Components as:

- the cost of employing people who are working within the Working Areas;
- the cost of Equipment used within the Working Areas;
- the cost of Plant and Materials including delivery to and from the Working Areas;
- the cost of charges paid by the Contractor for provision and use in the Working Areas;
- the cost of manufacture and fabrication of Plant and Materials which are designed specifically for the Works and manufactured or fabricated outside the Working Areas;
- the cost of design of the Works and Equipment done outside the Working Areas.

In Options A and B, this is a consideration in pricing compensation events; in Options C, D and E, it is a consideration in pricing compensation events, and also in assessing the Contractor's payments.

Any other costs not included within the Schedule of Cost Components and Shorter Schedule of Cost Components, including overheads and also profit, is deemed to be included in the Fee.

5.13 Schedules of Cost Components

The contract contains two Schedules of Cost Components (SCCs) and they have two uses:

1. Under Options A, B, C, D and E, they define the cost components, which are included in assessment of compensation events.
2. Under Options C, D and E, they define the cost components for which the Contractor will be directly reimbursed.

The Schedules of Cost Components do not apply to Option F (Management Contract) where Defined Cost is the amount of payments to Subcontractors plus the cost of any work carried out by the Contractor himself.

For Options A and B the Shorter Schedule of Cost Components is a complete statement of the cost components under the definition of Defined Cost.

For Options C, D and E the Defined Cost includes payments due to Subcontractors in addition to the cost components in the SCC, for which work is not subcontracted.

The Schedule of Cost Components or the Shorter Schedule of Cost Components?

There are two Schedules of Cost Components in the contract. This is because there are differing uses for Defined Cost, dependent on which Main Option is chosen:

Options A &B   Defined Cost is only used for assessing compensation events, therefore the Shorter Schedule of Cost Components is used.

Options C, D & E Defined Cost is used for assessing compensation events, and for assessing the PWDD, therefore the Schedule of Cost Components is used, though the Shorter Schedule of Cost Components can be used by agreement between the Project Manager and the Contractor for assessing compensation events.

Option F   Defined Cost is the amount of payment due to Subcontractors and the PWDD by the Contractor himself therefore neither the Schedule of Cost Components nor Shorter Schedule of Cost Components is used.

The Schedule of Cost Components

The Schedule of Cost Components only applies when Option C, D or E is used. References to the Contractor do not include his Subcontractors.

**Differences between the Schedule of Cost Components
and the Shorter Schedule of Cost Components**

Schedule of Cost Components	Shorter Schedule of Cost Components

People

Reference to the Contractor means the Contractor and not his Subcontractors.	Reference to the Contractor means the Contractor and his Subcontractors.
Detailed list of people costs classified as, Defined Cost.	No detailed list, but specifically includes amounts paid by the Contractor for meeting the requirements of the law and for pension provision.

Equipment

Detailed list of Equipment costs classified as Defined Cost.	Reference to the published list identified within Contract Data Part 2.

Plant and Materials

No difference.

Charges

Detailed list of Charges classified as Defined Cost.	No detailed list.
Overhead costs calculated by reference to the percentage for Working Areas overheads.	Overhead costs calculated by reference to the percentage for people overheads.

Manufacture and fabrication

Reference to hours worked multiplied by hourly rates in Contract Data Part 2.	Reference to amounts paid by the Contractor.

Design

No difference.

Insurance

No difference.

Figure 5.2 Differences between the Schedule of Cost Components and the Shorter Schedule of Cost Components

Clause 1: People

This relates to the cost of people who are directly employed by the Contractor and whose normal place of working is within the Working Areas, ie the Site and any another area named in Contract Data Part 2, and also people whose normal place of working is not within the Working Areas but who are working in the Working Areas, ie people who are based off Site, but who are on Site temporarily.

The cost component covers the full cost of employing the people including wages and salaries paid whilst they are in the Working Areas, payments made to the people for bonuses, overtime, sickness and holiday pay, special allowances, and also payments made in relation to people for travel, subsistence, relocation, medical costs, protective clothing, meeting the requirements of the law, a vehicle and safety training.

NEC3 contracts do not provide for people rates to be priced as part of the Contractor's tender, the principle being that the people component is real cost rather than a rate that has been forecast at tender stage months or even years before it is required to be used. This is a fairer method of establishing cost in that the risk is not the Contractor's, but in practice requires the Contractor if necessary to prove the cost of each operative, tradesman and member of staff, which can at times be laborious.

For this reason, many Employers include a schedule of rates to be priced by the Contractor at tender stage, these rates the being used for payments and compensation events where appropriate. These rates are also referred to during the tender assessment process.

Clause 2: Equipment

This relates to the cost of Equipment (referred to as 'Plant' in other contracts) used within the Working Areas. If the Equipment is used outside the Working Areas, then it is deemed to be included in the Fee percentage.

If the Equipment is hired externally by the Contractor, the hire rate is multiplied by the time the Equipment is required. The cost of transport of Equipment to and from the Working Areas and any erection and dismantling costs are also costed separately.

If the Equipment is owned by the Contractor, or hired by the Contractor from a company within the parent company such as an internal 'plant hire' company, the cost is assessed at open market rates (not at the rate charged by the hirer) multiplied by the time for which the Equipment is required.

Difficulty has often arisen with past editions of the contract where a piece of Equipment is owned by the Contractor, or a company within the group, in which case there would be no invoice as such to prove the cost; in many cases it was just an internal charge or cost transfer.

In those previous editions, the drafters sought to deal with the problem by establishing the weekly cost by the use of a formula. The depreciation and maintenance charge is the actual purchase price of the item of Equipment divided by its working life remaining at purchase, then multiplied by the depreciation and maintenance percentage.

Example

- Purchase Price = £30,000
- Working life remaining = 250 weeks
- Depreciation and maintenance percentage = 20%

The weekly cost is:

$$\frac{£30,000 \times 1.20}{250} = £144.00 \text{ per week}$$

In theory, this should provide an equitable solution, but in practice it has provided some inequitable answers, and also it has to be applied to *every* piece of Equipment owned by the Contractor! One hopes he did not have too many of his own wheelbarrows on the Site!

In NEC3 the drafters then changed to 'open market rates multiplied by the time for which the Equipment is required'. In truth major contractors who have a large 'plant' company will benefit in this respect from advantageous discount agreements with their Suppliers which are well below 'open market rates'.

Equipment purchased for use in the contract is paid on the basis of its change in value (the difference between its purchase price and its sale price at the end of the period for which it is used) and the time related on cost charge stated in Contract Data Part 2 for the period the Equipment is required.

During the course of the contract, the Contractor is paid the time-related charge per time period (per week/per month) and when the change in value is determined, a final payment is made in the next assessment.

Example

The Contractor has purchased six Portakabins at £6,250 each for use on a project of four years' duration. This cost includes supply and delivery of the Portakabins.

- Total purchase Price = 6 × £6,250.00 = £37,500
- On Completion, the Portakabins are sold for £2,750 each, therefore the total sale Price is £16,500
- The change in value is therefore £37,500 − £16,500 = £21,000

Any special Equipment is paid on the basis of its entry in Contract Data Part 2.

Consumables such as fuel are also separately costed including any Materials used to construct or fabricate Equipment.

The cost of transporting Equipment to and from the Working Areas and the erection, dismantling and any modifying of the Equipment is costed separately.

Any people cost such as Equipment drivers and operatives involved with erection and dismantling of Plant should be included in the cost of people, not the Equipment they work on.

Clause 3: Plant and Materials

This deals with purchasing Plant and Materials, including delivery, providing and removing packaging and any necessary samples and tests. The cost of disposal of Plant and Materials is credited.

Clause 4: Charges

This covers various miscellaneous costs incurred by the Contractor such as temporary water, gas and electricity, payments to public authorities, and also payments for various other charges such as cancellation charges, buying or leasing of land, inspection certificates and facilities for visits to the Working Areas.

The cost of any consumables and Equipment provided by the Contractor for the Project Manager's or Supervisor's offices is also included as direct cost. (Note that the cost of the Contractor's own Equipment is covered in the Equipment section, and the Contractor's consumables are included within the Working Areas overheads percentage.)

Item 44 covers various Site consumables as listed from (a) to (j).

It is important to recognise that Item 44 only includes for provision and use of Equipment, supplies and services, but excludes accommodation and people, so if we consider the list (see Figure 5.3), it is important to recognise what is and is not included.

The cost is calculated by multiplying the Working Areas overheads percentage inserted by the Contractor in Contract Data Part 2 by the people cost items 11, 12, 13 and 14.

Clause 5: Manufacture and fabrication

This relates to the components of cost of manufacture or fabrication of Plant and Materials outside the Working Areas. Hourly rates are stated in Contract Data Part 2 for the categories of employees listed.

Clause 6: Design

This deals with the cost of design outside the Working Areas. Again, hourly rates are stated in Contract Data Part 2 for the categories of employees listed.

Schedule of Cost Components			
Item 44	**Included**	**Not included**	**Where is it included**
Catering	Catering Equipment Food Electricity bills	Canteen accommodation Catering staff	Equipment People
Medical facilities and first aid	First aid box and contents Stretchers	First aid personnel First aid accommodation	People Accommodation
Recreation	Television Radio Gym facilities		
Sanitation	Toilet rolls Hand towels Soap Barrier creams	Toilet accommodation	Equipment
Security	Security systems Alarms	Security personnel Guard dogs	People Equipment
Copying	Photocopier Photocopier toner and paper	Other stationery	Equipment
Telephone, telex, fax, radio and CCTV	Telephones Fax machine Mobile radios Connection charges Call charges		
Surveying and setting out	Surveying Equipment Surveyors	Site Engineer People	People
Computing	Hardware Software Printers Print paper	IT personnel	People
Hand tools not powered by compressed air	Electric tools		

Figure 5.3 The Schedule of Cost Components

Clause 7: Insurance

The cost of events for which the Contractor is required to insure and other costs to be paid to the Contractor by insurers are deducted from cost.

The Fee

Whilst many practitioners believe that the Fee simply covers 'overheads and profit', the definition is a little wider. All components of cost not listed in the Schedules of Cost Components are covered by the Fee percentage.

There is no schedule of items covered by the Fee percentage, but the following list, whilst not exhaustive, gives some examples of cost components not included in the Schedules of Cost Components:

- profit;
- the cost of offices outside the Working Areas, eg the Contractor's head office;
- insurance premiums;
- performance bond costs;
- corporation tax;
- advertising and recruitment costs;
- sureties and guarantees required for the contract;
- some indirect payments to staff.

From this one can see that there are some elements covered by the Fee percentage which could be mistakenly assumed as being covered by the Schedules of Cost Components.

Previous editions of the ECC had only one Fee percentage applicable to total cost, but the NEC3 ECC allows the Contractor to tender two Fee percentages:

1. the *subcontracted Fee percentage* applied to the Defined Cost of subcontracted work;
2. the *direct Fee percentage* applied to the Defined Cost of other work.

It is essential that these two Fee percentages are correctly allocated to the appropriate costs when assessing payments and compensation events.

In reality, many Contractors tend to bracket the two Fee percentages together as a single Fee percentage, which they are entitled to do, and this in effect makes life easier for the Contractor and the Project Manager when assessing payments and compensation events.

Note that under Options A and B, Defined Cost is the cost of components in the Shorter Schedule of Cost Components whether work is subcontracted or not, whereas under Options C, D and E, where the Schedule of Cost Components is normally used, reference to the Contractor means the Contractor and not his Subcontractors. Subcontractors' costs are dealt with separately.

The Shorter Schedule of Cost Components

The Shorter Schedule of Cost Components is restricted to the assessment of compensation events under Options A and B, but if the Project Manager and Contractor agree it may be used for assessing compensation events under Options C, D and E.

Clause 1: People

This relates to the cost of people who are directly employed by the Contractor and whose normal place of working is within the Working Areas, ie the Site and any another area named in Contract Data Part 2, and also people whose normal place of working is not within the Working Areas but who are working in the Working Areas, ie people who are based off Site, but who are on Site temporarily. It also includes people who are not directly employed by the Contractor but are paid according to the time worked whilst they are in the Working Areas, eg security guards, cleaners, etc.

Whilst there is not a definitive list of costs included as with Items 11, 12 and 13 of the Schedule of Cost Components, it includes amounts paid by the Contractor including those for meeting the requirements of the law and pension provision.

Again, although NEC3 contracts do not provide for people rates to be priced as part of the Contractor's tender, many Employers include a schedule of rates to be priced by the Contractor at tender stage, these rates then being used for payments and compensation events where appropriate. These rates are also referred to during the tender assessment process.

Clause 2: Equipment

This relates to the cost of Equipment used within the Working Areas.

The cost of Equipment is calculated by reference to a published list, for example BCIS (Building Cost Information Service) Schedule of Basic Plant Charges or the CECA (Civil Engineering Contractors Association) Dayworks Schedule. In Contract Data Part 2, the Contractor names the published list to be used and also inserts a percentage for adjustment against items of Equipment in the published list.

The Contractor also inserts rates into Contract Data Part 2 for Equipment not included within the published list. Any Equipment required which is not in the published list or priced within Contract Data Part 2 is then priced at competitively tendered market rates.

The time the Equipment is used is as referred to in the published list, which may have hourly, weekly or monthly rates.

By referring to the published list, whether the Equipment is then owned or hired by the Contractor is irrelevant.

The cost of transporting Equipment to and from the Working Areas and the erection and dismantling of the Equipment is costed separately. The cost of Equipment operators is included within the people costs.

Any Equipment not included within the published lists is priced at competitively tendered or open market rates.

Clause 3: Plant and Materials

This is exactly the same as for the Schedule of Cost Components and also deals with purchasing, delivery, providing and removing packaging and any necessary samples and tests. The cost of disposal of Plant and Materials is credited.

Clause 4: Charges

This covers various miscellaneous costs incurred by the Contractor such as temporary water, gas and electricity payments to public authorities, and also payments for various other charges, which may or may not be relevant dependent on the project.

These costs are not calculated on a direct cost basis but by reference to the percentage for people overheads which is applied to people costs.

Clause 5: Manufacture and fabrication

This relates to the components of cost of manufacture or fabrication of Plant and Materials outside the Working Areas. The calculation is based on amounts paid by the Contractor.

Clause 6: Design

This is exactly the same as for the Schedule of Cost Components and deals with the cost of design outside the Working Areas. Again, hourly rates are stated in Contract Data Part 2 for the categories of employees listed.

Clause 7: Insurance

The cost of events for which the Contractor is required to insure and other costs to be paid to the Contractor by insurers are deducted from cost.

5.14 Disallowed Cost

There are different clauses covering Disallowed Cost dependent on the selected Main Option.

Options C, D and E: Clauses C11.2(25), D11.2(25), E11.2(25)

Disallowed Cost is cost which the Project Manager decides:

*   *is not justified by the Contractor's accounts and records*
    The Contractor is obliged to keep accounts, proof of payments, communications regarding payments, and any other records as stated in the Works Information (Clauses 52.3 and 52.3). If he does not have the accounts and records to prove a cost, then the cost must be disallowed. It is not sufficient to merely prove that the goods or Materials are on the Site for the Project Manager to see, or held off Site at a designated place for inspection, as with many other contracts.

*   *should not have been paid to a Subcontractor or Supplier in accordance with his contract*

The Contractor is obliged to submit the proposed conditions of contract for each Subcontractor to the Project Manager for acceptance, unless an NEC contract is proposed, or the Project Manager has agreed that no submission is required. In addition, under Options C, D, E and F, the Contractor is required to submit the proposed Contract Data for each subcontract if an NEC contract is proposed and the Project Manager instructs the Contractor to make the submission. If the Contractor pays a Subcontractor or Supplier an amount that is not in accordance with his contract with them, then this is disallowed.

- *was incurred only because the Contractor did not*

    - *follow an acceptance or procurement procedure stated in the Works Information*
    The Works Information may require the Contractor to comply with a specific procedure in respect of, for example, submission of design proposals or procurement of Subcontractors. If the Contractor does not comply with the procedure then associated costs are disallowed.
    - *give an early warning which the contract required him to give*
    If the Contractor incurs a cost that could have been avoided if the Contractor had given early warning, then it is disallowed.

and the cost of

- *correcting Defects after Completion*
Correction of Defects after Completion is a Disallowed Cost. However, correction of Defects before Completion is not a Disallowed Cost. This is often a contentious issue with Employers and Project Managers objecting to paying the Contractor for correcting his own Defects!

    However, one must consider this clause logically. The Contractor has quite possibly priced the risk of having to correct Defects which are his liability within his original tender; on an Option C contract, for example, this Price will be included within the target. Option C is a cost reimbursable contract until Completion, and the only way the Contractor can be paid for this risk, should it materialise, is through the Defined Cost process as Option C is a cost reimbursable contract until Completion. It is not a compensation event, so the target and/or the Completion Date are not changed. The Contractor is correcting the Defect in his own time and potentially reducing the share he could be awarded at Completion.

- *correcting Defects caused by the Contractor not complying with a constraint on how he is to Provide the Works stated in the Works Information*
A constraint may be stated in the Works Information, such as a prescribed method of working. For example, the Works Information may prescribe that a hardcore sub-base filling must be rolled six times with a vibrating roller of a certain weight, but the Contractor does not do so, and later a Defect arises due to the Contractor's failure to comply with that stated constraint.

If the Contractor does not comply with this constraint and a Defect occurs, then the Contractor's cost of correcting the Defect is disallowed.

- *Plant and Materials not used to Provide the Works (after allowing for reasonable wastage) unless resulting from a change to the Works Information*
  Excess wastage of Plant or Materials beyond what is considered reasonable is a Disallowed Cost. The question of what is 'reasonable' can often be debatable! As with all Disallowed Cost, it is the Project Manager's responsibility to make the decision and to disallow the cost.

- *resources not used to Provide the Works (after allowing for reasonable availability and utilisation) or not taken away from the Working Areas when the Project Manager requested*
  This provision will include people and Equipment. If the Contractor does not remove Equipment when it is no longer required, then this is Disallowed Cost. If the Contractor is using more resources than he has planned and priced for, or his resources are inefficient, that is not a Disallowed Cost.

- *preparation for and conduct of an adjudication or proceedings of the tribunal*
  If an adjudication, arbitration or legal proceedings occur, then each party bears his own costs. This bullet point was not introduced until NEC3 was published, so prior to NEC3 a Contractor could, in theory, refer a dispute to adjudication and whether or not he was successful, any costs arising could be included as cost and would not be disallowed.

Option F: Clause F11.2(26)

Disallowed Cost is cost which the Project Manager decides:

- *is not justified by the Contractor's accounts and records*
  The Contractor is obliged to keep accounts, proof of payments, communications regarding payments, and any other records as stated in the Works Information (Clauses 52.3 and 52.3). If he does not have the accounts and records to prove a cost, then the cost must be disallowed. It is not sufficient to merely prove that the goods or Materials are on the Site for the Project Manager to see, or held off Site at a designated place for inspection, as with many other contracts.

- *should not have been paid to a Subcontractor or Supplier in accordance with his contract*
  The Contractor is obliged to submit the proposed conditions of contract for each Subcontractor to the Project Manager for acceptance, unless an NEC contract is proposed, or the Project Manager has agreed that no submission is required. In addition, under Options C, D, E and F, the Contractor is required to submit the proposed contract data for each subcontract if an

NEC contract is proposed and the Project Manager instructs the Contractor to make the submission. If the Contractor pays a Subcontractor or Supplier an amount that is not in accordance with his contract with them, then this is disallowed.

- *was incurred only because the Contractor did not*

    - *follow an acceptance or procurement procedure stated in the Works Information*
    The Works Information may require the Contractor to comply with a specific procedure in respect of, for example, submission of design proposals or procurement of Subcontractors. If the Contractor does not comply with the procedure then associated costs are disallowed.
    - *give an early warning which the contract required him to give*
    If the Contractor incurs a cost that could have been avoided if the Contractor had given early warning, then it is disallowed.

- *is a payment to a Subcontractor for*

    - *work which the Contractor states that the Contractor will do himself; or*
    - *the Contractor's management.*

Clearly, with all the Disallowed Cost provisions, as the Project Manager is to decide that a cost is to be disallowed, then he has to be diligent enough to identify the cost, to quantify it, and to make the appropriate deductions, which are not the easiest of tasks!

5.15   Project Bank Accounts

In 2008, the Office of Government Commerce (OGC) published a guide to fair payment practices, following which the NEC Panel prepared a document in June 2008 to allow users to implement these fair payment practices in NEC contracts.

This document entitled 'Z3: Project Bank Account' is introduced to the ECC by means of a Z clause and authorises a project bank account which receives payments from the Employer which is in turn used to make payments to the Contractor and Named Suppliers.

There is also a Trust Deed between the Employer, the Contractor and Named Suppliers containing the necessary provisions for administering the Project Bank Account. This is executed before the first assessment date.

The Contractor also includes in his subcontracts for Named Suppliers to become party to the Project Bank Account through a Trust Deed. The Contractor notifies the Named Suppliers of the details of the Project Bank Account and the arrangements for payment of amounts due under their contracts. The Named Suppliers will be named within the Contractor's tender, but also additional Named Suppliers may be included subject to the Project Manager's acceptance by means of a Joining Deed, which is executed by the Employer, the Contractor and the new

Named Supplier. The new Named Supplier then becomes a party to the Trust Deed.

As the Project Bank Account is maintained by the Contractor, he pays any bank charges and also is entitled to any interest earned on the account. The Contractor is also required at tender stage to put forward his proposals for a suitable bank or other entity which can offer the arrangements required under the contract.

The process every month is that at each assessment date, the Contractor submits an application for payment to the Project Manager including details of amounts due to Named Suppliers in accordance with their contracts.

No later than one week before the final date for payment, the Employer makes Payment to the Project Bank Account of the amount which is due to be paid to the Contractor. If the Project Bank Account has insufficient funds to make all the payments, particularly to the Named Suppliers, the Contractor is required to add funds to the account to make up the shortfall.

The Contractor then prepares the Authorisation, setting out the sums due to Named Suppliers. After signing the Authorisation, the Contractor submits it to the Project Manager for signature by the Employer and submission to the Project Bank.

The Contractor and Named Suppliers then receive payment from the Project Bank Account of the sums set out in the Authorisation after the Project Bank Account receives payment.

In the event of termination, no further payments are made into the Project Bank Account.

6 Compensation events

6.1 Introduction

Compensation events, including their definition, notification, quotations, assessment and implementation are covered within the ECC by Clauses 60.1(1) to 65.2 with further references within the Main Option clauses.

When considering the principle of compensation events under the ECC, it is appropriate first to consider how change is managed with most other standard forms of contract.

6.2 The 'traditional approach' to managing change

The 'traditional approach' to managing change in construction contracts is normally dealt with in three separate elements within the contract:

(i) Variation

The changes in the form of additions, omissions or substitutions are priced by measuring the changed work, then pricing the change using rates consistent with the Contractor's original tender, the pricing document for that tender normally being in the form of a bill of quantities, schedule of rates or a contract sum analysis.

Where the changed work is similar to that referred to in the pricing document, executed under similar conditions, and does not significantly change the quantity of work set out in the tender, the original rates and prices apply. Where the changed work is similar to that referred to in the tender, but not executed under similar conditions, and/or there is a significant change in the quantity of work, the original rates and prices form the basis of the pricing, ie they are adjusted to allow for the differences in conditions and/or quantities. There are normally provisions for the parties to agree fair rates and prices and where the work cannot be clearly defined and/or measured then daywork provisions apply.

Whilst this has always been seen as the 'normal' way to price change in a construction contract, it has its drawbacks.

The first problem with the traditional approach is that the rates and Prices inserted by the Contractor in his tender were calculated based on the original quantities and whether they are right or wrong they are being applied to changes which are outside his control. In that sense, either the Contractor or the Employer may gain or lose financially by the changes.

The second problem is that there are rarely specific time scales within the contracts for submitting the prices and agreeing them, apart from the Final Account which is normally submitted by the Contractor and agreed by the contract administrator, months or even years after the Works have been completed.

(ii) Extension of time

Whilst the variation element deals with the changes to the prices, the effect on the Completion Date is dealt with as a separate process referred to as 'extension of time' or 'adjustment to Completion Date'. This provision allows for the Contract Administrator to extend time/fix a new Completion Date, preventing time becoming at large and thereby postponing the Employer's right to recover liquidated damages.

In that respect, the Contractor is required to submit a separate notice to the Contract Administrator upon it becoming apparent that a delay will occur. Normally there is a time scale within these provisions for the Contractor to submit his notice and in turn for the Contract Administrator to respond.

The purpose of the notice requirement is to allow the Contract Administrator to attempt to mitigate the problem either by stopping the delay from recurring or even omitting work in addition to making a judgement on the Contractor's entitlement to additional time.

The notice will normally identify:

* the cause of the delay – detail of how and why the delay is occurring or likely to occur;
* the estimated effect on Completion – the Contractor should also give an estimate of delay in his notice so that the Contract Administrator can form his own opinion;
* the clauses on which the Contractor relies in requesting an extension of time.

(iii) Loss and expense

Finally, contracts have provision for the Contractor to recover loss and expense which he could not have recovered under any other provision within the contract.

Loss and expense claims have from experience been given a very bad name. In reality loss and expense claims should be a process by which the Contractor can submit a well-structured, fully substantiated claim to seek to recover any loss or expense which he has incurred as a result of something which is outside his control and/or foreseeability.

Whilst the financial effect of a variation should essentially be priced with the variation, the cumulative effect of a number of variations or other factors which are not the Contractor's risk may involve the Contractor in loss and expense in having to work uneconomically, out of sequence, etc and this provision deals with that possibility. They are generally known as disruption claims. Often agreement to loss and expense claims is a prolonged process which extends long after Completion of the project.

6.3 The NEC3 approach

The NEC3 contracts take a very different approach to pricing and managing change as contained within the compensation event provisions, whereby the event is notified, priced, assessed and implemented within a defined and relatively short time scale, including the cost and time effect of the change.

It recognises that it is critical that the time and cost effect of a change are considered together as the result of that change, therefore the Prices and the Completion Date are changed within a single provision rather than within three separate processes as with the traditional approach previously described.

A very important difference to the traditional methods is that no reference is made to any of the Contractor's contract pricing except by the agreement of the parties. This ensures that the Contractor's original financial and cash flow structure is unaffected by the compensation event. All time and cost effects are contained within the evaluation of the compensation event.

The other important difference is that what circumstances constitute the basis of a compensation event are defined within the contract. This gives the

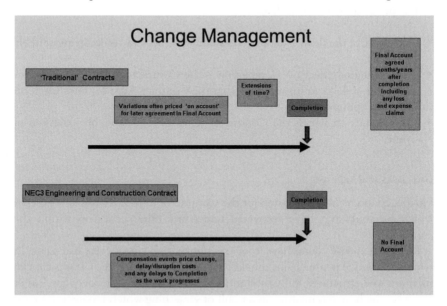

Figure 6.1 Comparison of change management provisions under 'traditional' contracts and the NEC3 Engineering and Construction Contract

contract drafter the facility to adjust the balance of risk between the Employer and Contractor.

Experience shows that generally the requirements of the compensation events procedures are poorly understood and applied. The more onerous requirements on the Project Manager introduced in NEC3 were included to meet a particular failing, as were the time bar requirements on both parties.

6.4 Definition of a compensation event

There is no definition of a compensation event stated within the contract; however it could be defined as 'an event which is the Employer's risk, and if it occurs, and affects the Contractor, entitles the Contractor to be compensated for any effect on the Prices and the Completion Date'.

6.5 The compensation events

The compensation events may be found in a number of locations within the contract:

- 19 events are stated in Sub-clauses 60.1(1) to (19) – these are core clauses and therefore always apply unless amended by Z clause provisions;
- a further three events are stated in Options B and D Sub-clauses 60.4 to 60.6 – these are the Options which are based on a bill of quantities prepared by the Employer and therefore if, for example, there is a mistake in the bill of quantities then a compensation event applies;
- further events are included in Secondary Options X2 (changes in the law), X12 (partnering), X14 (advanced payment to the Contractor), X15 (limitation of the Contractor's liability for his design to reasonable skill and care) and Y(UK)2;
- additional compensation events may also be stated by the Employer in Contract Data Part 1;
- also compensation events can be added, omitted or amended through Z clauses.

The principal compensation events are set down in Clause 60.1:

> *60.1(1) The Project Manager gives an instruction changing the Works Information except*
>
> - *a change made in order to accept a Defect or*
> - *a change to the Works Information provided by the Contractor for his design which is made either at his request or to comply with other Works Information provided by the Employer.*

Changes to the Works Information in the ECC are normally defined in other contracts as 'variations'. The ECC does not use the term 'variations', but the Project

Manager can give an instruction changing the Works Information, which can be in the form of an addition, omission or alteration to work.

An instruction can also be given in order to resolve an ambiguity or an inconsistency within the Works Information. An ambiguity may be defined as 'an uncertain meaning or intention' and therefore is something that is unclear; an inconsistency may be defined as a 'conflict in meaning or intention' and therefore could be a mismatch or discrepancy between documents.

An example of an inconsistency is that the drawings in the Works Information provided by the Employer show a particular grade of concrete, but the specification shows a different grade. Under Clause 17.1, the Project Manager or Contractor notifies the other as soon as either becomes aware of the ambiguity or inconsistency and the Project Manager then gives an instruction resolving it; the subsequent change to the Works Information is then a compensation event under Clause 60.1(1).

Under Clause 63.8, a compensation event which arises from an instruction to change the Works Information to resolve an ambiguity or inconsistency is assessed 'as if the Prices, the Completion Date and the Key Dates were for the interpretation most favourable to the party which did not provide the Works Information'. In this case it would be assessed in favour of the Contractor who is deemed to have priced the cheaper, easier or quicker alternative.

Note, from the first bullet point in Clause 60.1(1), that if the change to the Works Information is made in order to accept a Defect (Clause 44), then it is not a compensation event.

As detailed previously in Chapter 4, the Contractor and the Project Manager may each propose to the other that the Works Information should be changed so that a Defect does not have to be corrected, in effect nullifying the Defect. This would not be a compensation event.

Also, from the second bullet point, a change to the Works Information provided by the Contractor for his design which is made either at his request or to comply with other Works Information provided by the Employer is not a compensation event. So Works Information provided by the Employer in Part 1 of the Contract Data takes precedence over that provided by the Contractor in Part 2 of the Contract Data.

Example

The Works Information included in Contract Data Part 1 states that all doors are to have brass furniture. The Contractor states in his tender in Contract Data Part 2 that all doors will have aluminium furniture. The requirement for brass furniture in the Employer's Works Information in Contract Data Part 1 will prevail.

Thus it is critical that the Contractor ensures that the Works Information he prepares and submits with his tender as Part 2 of the Contract Data complies with the requirements of the Employer's Works Information in Part 1 of the Contract Data.

Also, under Clause 18.1 the Contractor notifies the Project Manager as soon as he considers that the Works Information requires him to do anything which is illegal or impossible, the parties cannot contract to do something illegal. If the Project Manager agrees he gives an instruction to change the Works Information appropriately.

> *60.1(2) The Employer does not allow access to and use of a part of the Site by the later of its Access Date and the date shown on the Accepted Programme.*

The Employer has a fundamental obligation to allow the Contractor access to the Site (Clause 33.1). This clause recognises two dates, that shown in Contract Data Part 1, and that shown in the Accepted Programme.

The Project Manager should be aware that if the Contractor shows a date in his programme when he requires access to and use of the Site or a part of the Site and the Project Manager accepts the programme, then the Employer is obliged to allow that access by the later of that date and the date in the Contract Data, failing which a compensation event occurs. So the Contractor can relax the obligation to the date in the contract by showing a later date in the Accepted Programme, but cannot impose an earlier date than that stated in the contract.

> *60.1(3) The Employer does not provide something which he is to provide by the date for providing it shown on the Accepted Programme.*

The Employer is obliged to provide information, facilities or other things as shown on the Accepted Programme. It is important that in addition to the Employer stating the dates when he will provide things, the Contractor shows these requirements and the dates for providing them in his programme submitted for acceptance, so that when the Project Manager accepts the programme he accepts liability on behalf of the Employer if he fails to provide it.

Note that, whilst the sub-clause refers to 'the Employer', it applies to anyone who acts for the Employer and has an obligation to provide something, so this clause could include, for example, the Employer not providing free-issue Materials to the Contractor, or the Project Manager not providing design information to the Contractor.

> *60.1(4) The Project Manager gives an instruction to stop or not to start any work or to change a Key Date.*

It is important to note that the Project Manager giving an instruction to stop or not to start any work or to change a Key Date is always a compensation event, regardless of the reason why the instruction was given; however, if the Project

Manager gives an instruction to the Contractor to stop a part of the work, for example because the Contractor is working unsafely and the Contractor notifies the Project Manager that he believes it is a compensation event, then under Clause 61.4, if the Project Manager decides that the event arose from a fault of the Contractor, then he notifies the Contractor that the Price, Completion Date and the Key Dates are not to be changed.

60.1(5) The Employer or Others

- *do not work within the times shown on the Accepted Programme,*
- *do not work within the conditions stated in the Works Information or*
- *carry out work on the Site that is not stated in the Works Information.*

The words 'do not work within the times shown on the Accepted Programme' can often cause debate. What work is the Employer, or Others, required to do? Also, 'work within' implies a process over a period of time rather than a specific task or date so it may be difficult to determine whether someone has worked within the time shown on the Accepted Programme.

Again, it is important that in addition to the Employer showing what he is going to do, the Contractor shows these requirements and the dates for providing them in his programme submitted for acceptance, so that when the Project Manager accepts the programme he accepts liability on behalf of the Employer if he fails to provide it.

Note that the term 'Others' are people who are not the Employer, Project Manager, Supervisor, Adjudicator, Contractor, Subcontractor or Supplier, examples of Others being utilities companies installing or diverting electricity, gas or water services, and is a traditional area of difficulty. Some Employers amend this clause to exclude their liability for 'Others' thus passing the risk to the Contractor. This is unreasonable because the Contractor usually has no more contractual or financial control over the service providers than the Employer.

60.1(6) The Project Manager or the Supervisor does not reply to a communication from the Contractor within the period required by this contract.

If the Project Manager or Supervisor does not reply to a communication, which could be, for example, the Contractor submitting the names of a proposed Subcontractor, design proposals or a programme, etc, then it is a compensation event as it is fundamentally a breach of contract. However, if the Contractor notifies a compensation event in this respect, and the Project Manager decides that event has no effect upon Defined Cost, Completion or meeting a Key Date, then again under Clause 61.4, he notifies the Contractor that the Price, Completion Date and the Key Dates are not to be changed.

The 'period required by this contract' may either be an expressly stated period of time, for example two weeks, for the Project Manager to reply to the submission of programme, or the more general 'period for reply' stated in Contract Data Part 1.

> *60.1(7) The Project Manager gives an instruction for dealing with an object of value or of historical or other interest found within the Site.*

The term 'object of value or of historical or other interest' could be very wide ranging and subject to interpretation; for example, what value is classed as 'of value' – would a mobile phone or a watch be something of value? How old does something have to be to be classed as 'historical'? The key is that the Project Manager gives an instruction for dealing with the object, which is unlikely if a personal possession such as a watch was found. The clause normally deals with buried articles such as fossils, archaeological remains, but can also include munitions or other wartime remains. Such finds may also be governed by government legislation, local bylaws, etc and may involve investigation by educational establishments, museums and other authorities.

Note that under Clause 73.1, the Contractor has no title to objects of value or historical interest within the Site, but notifies the Project Manager when such an object is found and the Project Manager instructs him how to deal with it.

> *60.1(8) The Project Manager or the Supervisor changes a decision which he has previously communicated to the Contractor.*

The contract refers to 'decisions' under the following clauses:

- Clause 11.2(25) – the Project Manager decides Disallowed Cost.
- Clause 30.2 – the Project Manager decides the date of Completion.
- Clause 50.1 – the first assessment date is decided by the Project Manager.
- Clause 61.4 – the Project Manager decides whether an event notified by the Contractor is a compensation event.
- Clause 61.5 – the Project Manager decides that the Contractor did not give an early warning of the event which an experienced Contractor could have given.
- Clause 61.6 – the Project Manager decides that the effects of a compensation event are too uncertain to be forecast reasonably.
- Clause 63.5 – the Project Manager has notified the Contractor of his decision that the Contractor did not give an early warning of the event which an experienced Contractor would have given.
- Clause 64.1 – the Project Manager assesses a compensation event if he decides that the Contractor has not assessed the compensation event correctly.

This clause has been interpreted by some NEC3 practitioners as a provision to allow the Project Manager to change and revise his assessment of an implemented compensation event. This is an incorrect interpretation; it must be stressed that the Project Manager does not have the authority to change his assessment once it has been implemented. The Project Manager's assessment is final and the only recourse would be adjudication.

> *60.1(9) The Project Manager withholds an acceptance (other than acceptance of a quotation for acceleration or for not correcting a Defect) for a reason not stated in this contract.*

The core clauses dealing with acceptance are:

- Clause 15.1 – acceptance of a proposal for adding to the Working Areas;
- Clauses 21.2/21.3 – acceptance of the Contractor's design;
- Clause 23.1 – acceptance of the Contractor's design of Equipment;
- Clause 24.1 – acceptance of the Contractor's replacement key person;
- Clause 26.2 – acceptance of a proposed Subcontractor;
- Clause 26.3 – acceptance of the proposed conditions of contract for a Subcontractor;
- Clause 31.3 – acceptance of a programme;
- Clause 44.2 – acceptance of a quotation to accept a Defect;
- Clause 62.3 – acceptance of a Contractor's quotation for a compensation event;
- Clause 85.1 – acceptance of the Contractor's insurance certificates.

Additionally, under Options C, D, E and F:

- Clause 26.4 – acceptance of the proposed Contract Data for a subcontract.

Additionally, within the Secondary Options:

- X13 – acceptance of the performance bond;
- X14 – acceptance of the advanced payment bond.

Within each of the clauses are stated the reasons for not accepting. If the Project Manager does not accept for a different reason then this is a compensation event. For example, the Project Manager may not accept a proposed Subcontractor as he has never heard of them. The Contractor may offer additional information about this proposed Subcontractor, but if the Project Manager still does not accept them as he has never heard of them, then that is not that the appointment of the Subcontractor 'will not allow the Contractor to Provide the Works'.

> *60.1(10) The Supervisor instructs the Contractor to search for a Defect and no Defect is found unless the search is needed only because the Contractor gave insufficient notice of doing work obstructing a required test or inspection.*

This relates to the Supervisor instructing the Contractor to search for a Defect under Clause 42.1. Note that if the Contractor did not give sufficient notice of doing work obstructing a test or inspection then it is not a compensation event, so, for example, if the Contractor backfilled drain trenches before the Supervisor had reasonable opportunity to inspect them, or the contract specifically stated a period for inspection and the Contractor did not allow that period, then this would be the Contractor's risk and not a compensation event.

60.1(11) A test or inspection done by the Supervisor causes unnecessary delay.

Under Clause 40.5, the Supervisor does his tests without causing unnecessary delay to the work, so he needs to act and carry out that test or inspection in a timely fashion. How long the test or inspection is delayed before it moves from 'reasonable' to 'unreasonable' delay is clearly a subjective judgement, but if the Supervisor delays the Contractor beyond what would be deemed 'necessary' bearing in mind the type of test or inspection, then it is a compensation event.

60.1(12) The Contractor encounters physical conditions which

- *are within the Site,*
- *are not weather conditions and*
- *an experienced contractor would have judged at the Contract Date to have such a small chance of occurring that it would have been unreasonable for him to have allowed for them.*

Only the difference between the physical conditions encountered and those for which it would have been reasonable to have allowed is taken into account in assessing a compensation event.

It is important to interpret this final paragraph correctly because if the Contractor did not allow anything in terms of Price and/or time within his tender for dealing with a physical condition, he should only be compensated for the difference between what he found and what he should have allowed, so the Contractor is unlikely to be compensated for the full value of dealing with the physical condition. If he should have allowed time and/or money within his tender for dealing with a physical condition then this must be considered in assessing a compensation event.

Note that, under Clause 60.2, in judging the physical conditions for a compensation event, the Contractor is assumed to have taken into account:

- the Site Information – this could include Site investigations, borehole data, etc;
- publicly available information referred to in the Site Information – this could include reference to public records about the Site;
- information obtainable from a visual inspection of the Site – note the use of the term 'visual', the Contractor is not assumed to have carried out an intrusive investigation of the Site;

- other information which an experienced Contractor could reasonably be expected to have or to obtain – this is a fairly subjective criterion, but precludes the Contractor from relying solely on what is contained within the Site Information. Note that the clause refers to 'an experienced Contractor', not 'the Contractor' or anyone who has a detailed local knowledge of the Site.

Under Clause 60.3, 'if there is an ambiguity or inconsistency within the Site Information (including the information referred to in it), the Contractor is assumed to have taken into account the physical conditions more favourable to doing the work', this could be the cheaper, easier or quicker alternative.

This complies with the rule of 'contra proferentem' (contra = against, proferens = the one bringing forth) in that where a term or part of a contract is ambiguous or inconsistent it is construed strictly against the party which imposes or relies on it.

It is critical that the Employer makes all information in his possession available to the Contractor. He cannot be selective, withholding information with a view to obtaining advantage.

It is also important to note that if the Contractor encounters unforeseen physical conditions, which are often but not always, ground conditions, he may not necessarily be compensated for the cost and time effect of dealing with it, so he must consider carefully the wording of Clauses 60.1(12) and 60.2 to prove his case.

It is also important to recognise that compensation to the Contractor is assessed as the difference between what the Contractor found, and what it would have been reasonable for him to have allowed in his tender, not simply the difference between what he found and what he did allow in his tender.

60.1(13) A weather measurement is recorded

- *within a calendar month,*
- *before the Completion Date for the whole of the Works and*
- *at the place stated in the Contract Data*

the value of which, by comparison with the weather data, is shown to occur on average less frequently than once in ten years.

Only the difference between the weather measurement and the weather which the weather data show to occur on average less frequently than once in ten years is taken into account in assessing a compensation event.

With regard to weather-related delays, most contracts use the words 'exceptionally adverse weather' or 'exceptionally adverse climatic conditions', leaving it to the parties to determine, and in many cases to argue, what is meant by the words 'exceptionally adverse'. 'Exceptionally' may be defined as 'more than normal' whilst 'adverse' may be defined as 'unfavourable' in the sense that it impacted upon the Contractor's progress, which has to take into account the weather that would reasonably have been expected bearing in mind the location of the Site, the time of year and what the Contractor was intending to do at the time the weather condition occurred.

The ECC provides a more objective approach by referring and comparing to weather data provided by an independent party such as a meteorological office, an airport or military base.

Only the difference between the weather measurement and the weather which the weather data show to occur less frequently than once in ten years is taken into account in assessing a compensation event, so the Contractor will not be compensated for the full impact of the weather event as weather likely to occur within a ten-year period is the Contractor's risk, and it is assumed that the Contractor has already allowed for that in terms of Price and programme.

Contract Data Part 1 defines the place where the weather is to be recorded. Note that Clause 60.1(13) states 'at the place stated in the Contract Data' so it is vital that the historic weather data and the current weather measurements are recorded at the same place. It is important that the place chosen will truly reflect the likely conditions which will be encountered at the Site.

It also lists weather measurements for each calendar month in respect of:

- the cumulative rainfall (mm);
- the number of days with rainfall more than 5 mm – whilst location and time of year is clearly a factor, from analysis of meteorological records, 5 mm rainfall in a day is a fairly low level;
- the number of days in the month with minimum air temperature less than 0 degrees Celsius; and
- the number of days in the month with snow lying at a stated time GMT – there is no measure of how much snow, merely that it is lying, presumably on the ground at a stated time.

There is also provision for adding other measurements which could include wind speed and other weather-related data. This might apply at, say, coastal or mountain locations.

Note also that the weather measurement is recorded before the Completion Date for the whole of the Works and at the place stated in the Contract Data.

It has to be said that, although the NEC drafters have recognised that weather and its effects should be more objectively measured and managed, there still remains some confusion as to when a weather-related compensation event exists and how to assess it.

Ironically, it is even more important to assess the event correctly with ECC than with other contracts, as the Contractor is awarded time as well as money, whereas other contracts tend to only award time.

As a compensation event under Clause 60.1(13) occurs when a weather measurement is recorded within a calendar month, the value of which, by comparison with the weather data, is shown to occur on average less frequently than once in ten years, there is a 'trigger point' in the month at which the weather, for example the cumulative rainfall, exceeds the 'once in ten years' test. Note that the compensation event occurs once the trigger point has been exceeded, not once it has an effect on the progress of the Works.

The weather up to the trigger point is at the Contractor's risk in terms of programme and cost and is deemed to have been included within the Contractor's tender, but once the trigger point is reached, there are two schools of thought as to how the compensation event operates:

1. If the trigger point is reached on, say, the 25th of the month, only delays incurred by the weather *after* the 25th are compensated for.
2. If the trigger point is reached on, say, the 25th of the month, one must consider the *whole* month, but only the difference between the weather measurement and the weather which the weather data show to occur on average less frequently than once in ten years is taken into account in assessing a compensation event.

In either case, one must consider the worst weather in ten years, not the average weather. Also, once a compensation event exists, it is incumbent upon the Contractor to submit a quotation showing any effect that the weather had on the Prices and/or any delay to Completion. There is no automatic entitlement to be paid for the weather merely because the trigger point was passed.

60.1(14) An event which is an Employer's risk stated in this contract.

The Employer's risks are stated in Clause 80.1 but additional Employer's risks may also be inserted into Contract Data Part 1.

There are six main categories of Employer's risk in the contract and therefore covered by this clause:

1. risks relating to the Employer's use or occupation of the Site;
2. risks relating to items supplied by the Employer to the Contractor until the Contractor has received them, or up to the point of handover to the Contractor;
3. risks relating to loss or damage to the Works, Plant and Materials, caused by matters outside the control of the parties;
4. risks arising once the Employer has taken over completed work, except a Defect that existed at take-over, an event which was not an Employer's risk, or due to the activities of the Contractor on Site after take-over;
5. risks relating to loss or wear or damage to parts of the Works taken over by the Employer, and any Equipment, Plant and Materials retained on Site after termination;
6. any other risks referred to in the Contract Data. These should be stated in Contract Data Part 1.

60.1(15) The Project Manager certifies take-over of a part of the Works before both Completion and the Completion Date.

Rainfall — July (mm)

Date	1	2	3	4	5	6	7	8	9	10	11	12	13	14	15	16	17	18	19	20	21	22	23	24	25	26	27	28	29	30	31	mm
Daily Rainfall	2	3	2	4	2	2	3	7	8	8	7	2	3	2	3	4	4	2	2	3	2	3	2	3	2	3	4	2	2	2	2	100
Cum. Rainfall	2	5	7	11	13	15	18	25	33	41	48	50	53	55	58	62	66	68	70	73	75	78	80	83	85	88	92	94	96	98	100	
Worst Rainfall																								80								80

Period of delay (shaded between dates 8–10)

Period of delay (shaded around date 16)

Compensation event triggers on 24th

Principle 1 Compensation event triggers on 24th, only weather delays after that date are assessed.

Principle 2 Compensation event triggers on 24th, consider the whole of the month versus the month falling within the more frequently than 1 in 10 years range.

Figure 6.2 Weather-related compensation events

Under Clause 35.2, the Employer may use part of the Works before Completion has been certified, in which case he takes over the part of the Works when he begins to use it, except if the use is for a reason stated in the Works Information or to suit the Contractor's method of working. If take-over occurs before the Contractor has completed and the Completion Date, it is a compensation event.

If the Employer takes over part of the Works before Completion but after the Completion Date, eg because the Contractor is late completing and the Employer wants to take over part of the Works, then it is not a compensation event.

60.1(16) The Employer does not provide Materials, facilities and samples for tests and inspections as stated in the Works Information.

Under Clause 40.2, the Contractor and the Employer provide Materials, facilities and samples for test and inspections in accordance with the Works Information. If the Employer does not comply in providing the facilities, or is late in providing them, it is a compensation event.

60.1(17) The Project Manager notifies a correction to an assumption which he has stated about a compensation event.

Under Clause 61.6, if the Project Manager decides that the effects of a compensation event are too uncertain to be forecast reasonably, he states assumptions, the quotation being based on these assumptions (see example below). If the Project Manager later notifies a correction to the assumption, it is a compensation event. In this case, the change to the Prices and any delay to Completion the Contractor had assumed in his original quotation are then replaced with corrected amounts.

Example

The Project Manager instructs the Contractor to assume the cost of £5,000 for a part of a quotation for a compensation event which is too uncertain to be forecast reasonably. The cost of this element is subsequently recorded by the Contractor and found to be £6,200. The Project Manager notifies a correction to the assumption, which is then priced as follows:

Cost from records	£6,200
Project Manager's Assumption	£5,000
	£1,200
Add Fee @ 10%	£120
Change to the Prices	**£1,320**

60.1(18) A breach of contract by the Employer which is not one of the other compensation events in this contract.

This is a 'catch all' to cover any breaches not covered by the other compensation events; an example could be the Employer has not paid a certificate and the effect of the non-payment is that the Contractor has to delay or stop progress.

> *60.1(19) An event which*
>
> - *stops the Contractor completing the Works or*
> - *stops the Contractor completing the Works by the date shown on the Accepted Programme,*
>
> *and which*
>
> - *neither party could prevent,*
> - *an experienced Contractor would have judged at the Contract Date to have such a small chance of occurring that it would have been unreasonable for him to have allowed for it and*
> - *is not one of the other compensation events stated in this contract.*

Clause 60.1(19) is a new 'no fault' clause introduced under NEC3, and dealing with an event neither party could prevent, which stops the Contractor completing the Works. It is also tied in to Clause 19.1 'Prevention' following which, if the event occurs, the Project Manager instructs the Contractor how to deal with the event.

There has been much confusion and debate regarding these clauses since NEC3 was first published in 2005.

The intention of the drafters appears to be that this would be a *force majeure* (superior force) clause, often referred to as an 'Act of God' in other contracts and legal documents, ie significant events such as flooding, earthquakes, volcanic eruptions and the associated dust clouds and other natural disasters which prevent the contracting parties from fulfilling their obligations under the contract, the clause essentially freeing them, or giving some relief from their liabilities.

Whilst this appears to be what the drafters had intended, on closer examination the application of this clause is far wider.

If you take each element in turn:

- *stops the Contractor completing the Works*
  Due to the event, the Works were never completed.

- *stops the Contractor completing the Works by the date shown on the Accepted Programme*
  Due to the event, Completion of the Works was delayed.

- *neither party could prevent*
  Clearly, if either party could have prevented the event, then they would or should have. The clause refers to preventing the event, rather than taking some mitigating action to lessen its effect.

- *an experienced Contractor would have judged at the Contract Date to have such a small chance of occurring that it would have been unreasonable for him to have allowed for it*
  At tender stage, the likelihood of the event was so low, the Contractor would not have allowed for it.

- *is not one of the other compensation events stated in this contract*
  If the matter could have been dealt with as one of the other compensation events, then it should be. 60.1(19) then clearly deals with the exceptional event.

Examples of where this clause would at least cause confusion:

- Clause 60.1(13) has been deleted from the contract by means of a Z clause; one would then assume that the risk of unforeseen weather conditions is the Contractor's and he should have allowed for it.
- The Site is affected by a severe storm which occurs less frequently than once in 50 years, which delays Completion of the Works by one week.
- The Contractor states that:

  - The weather stopped him completing the Works by the date shown on the Accepted Programme. Assume that the Contractor can prove that.
  - Neither party could prevent the event. Whilst one can take measures to lessen the impact of a storm, one cannot prevent the storm from occurring.
  - An experienced Contractor would have judged at the Contract Date to have such a small chance of occurring that it would have been unreasonable to have allowed for it. The Contractor states that he could not have predicted or allowed for such a rare event and its consequences.
  - It is not one of the other compensation events. If Clause 60.1(13) has been deleted from the contract by means of a Z clause, one could argue that it is no longer one of the other compensation events.

Could Clause 60.1(19) also extend to loss or damage occasioned by insurable risks such as fire, lightning, explosion, storm, flood, earthquake and terrorism? Although both Schedules of Cost Components require that the cost of events for which the contract requires the Contractor to insure should be deducted from cost, the Contractor could still recover any delay to Completion through the compensation event.

If Clause 60.1(19) is intended as a form of *force majeure* provision, rather than including a fairly long-winded definition within the clause, why not just state: 'A *force majeure* event occurs' and allow the parties to interpret this internationally recognised term as they do in other contracts?

6.6 Additional compensation events within Options B and D

Under Options B and D, the Options which use bills of quantities, there are three additional compensation events.

(i) Clause 60.4

A difference between the final total quantity of work done and the quantity for an item on the bill of quantities is not a compensation event in itself, unless it satisfies all three bullet points within the clause.

This clause has caused some confusion, and has even led practitioners to delete it as a Z clause as it is seen either as confusing or unfair in its application, but if we examine each of the bullet points in turn:

- *The difference does not result from a change to the Works Information.*
  If it had resulted from a change to the Works Information, then it would have been dealt with as a compensation event under Clause 60.1(1).

- *The difference causes the Defined Cost per unit of quantity to change.*
  The purchase Price of Materials may be affected by a change in the quantity, for example a Price reduction or increase may be applicable when the number of units exceeds or falls below a certain amount. Or the cost of mobilisation or delivery could have changed as the same vehicle delivers fewer units.

- *The rate in the bill of quantities for the item multiplied by the final total quantity of work done is more than 0.5 per cent of the Total of the Prices at the Contract Date.*
  This final bullet point compares the final extended Price for that item in the bills of quantities with the Total of the Prices at the Contract Date (many contracts refer to this as the 'Tender Sum' or the 'Contract Sum').

If the extended Price is more than 0.5 per cent of the Total of the Prices at the Contract Date, and both of the previous bullet points are satisfied, then it is a compensation event. This final bullet point excludes items of a 'minor value' in the bill of quantities.

Example

Original quantity of work

Road Gullies 150 @ £125 = £18,750

Due to an error in measuring the road gullies when preparing the bill of quantities, the final quantity is only 120 so

Final quantity of work

Road Gullies 120 @ £125 = £15,000

The Total of the Prices at the Contract Date is £2,000,000.

0.5% × £2,000,000 = £10,000, therefore, assuming that all three conditions are satisfied, then it is a compensation event.

(ii) Clause 60.5

A difference between the final total quantity of work done and the quantity for an item on the bill of quantities which delays Completion or the meeting of a Condition stated for a Key Date is a compensation event.

Again, a difference in quantity alone is not a compensation event, but if the difference also delays Completion of the whole of the project or the meeting of a Key Date, then it is a compensation event.

(iii) Clause 60.6

This clause covers correction of mistakes in the bill of quantities which are departures from the rules for measurement, or are due to ambiguities or inconsistencies.

It is the Employer's responsibility to properly include and describe items in the bills of quantities. If a mistake arises it is therefore dealt with as a compensation event. This will include items included within the Works Information which are not included within the bills of quantities.

Finally, under Clause 60.7, where there is an inconsistency between the bill of quantities and another document, eg the Works Information, the Contractor is assumed to have taken the bill of quantities as correct. Whilst the Works Information defines the work the Contractor has to do, the bill of quantities is the document on which he prices his tender and therefore he can safely assume that the bill of quantities correctly represents the work he has to price. Again, this reinforces the Employer's responsibility to properly include and describe items in the bills of quantities.

6.7 Notifying compensation events

Compensation events may either be notified by the Project Manager, or the Contractor can notify a compensation event to the Project Manager.

- The Project Manager notifies the Contractor of the compensation event at the time of giving the instruction (Clause 61.1). He also instructs the Contractor to submit quotations, unless the event arises from a fault of the Contractor or quotations have already been submitted. The Contractor puts the instruction or changed decision into effect.
- If the Project Manager instructs the Contractor to submit a quotation for a proposed instruction, the Contractor has only been instructed to submit a quotation; he does not carry out the work until instructed to do so.
- The Contractor can notify a compensation event, but must do so within eight weeks of becoming aware of the event, otherwise he is not entitled to a change in the Prices, the Completion Date or a Key Date, unless the Project Manager should have notified the event to the Contractor but did not (Clause 61.3).

The intention of the clause is to compel the Contractor to notify compensation events promptly, otherwise any entitlement to additional time and money is lost. Note that the eight-week rule requires the Contractor to notify the compensation event to the Project Manager within eight weeks of becoming aware of it, not just to have given early warning or mentioned the possibility of a compensation event in a discussion with the Project Manager.

The clause therefore covers compensation events initiated by the Contractor, rather than the Project Manager, examples being the weather and unforeseen ground conditions which the Project Manager may not have been aware of unless the Contractor had notified it.

An example of the Project Manager failing to notify a compensation event when he should have would be if an instruction was issued by the Project Manager to the Contractor changing the Works Information, but at the time the Project Manager did not notify the compensation event, and the Contractor in turn did not notify it either. Even if the compensation event is not notified within eight weeks, the responsibility remains with the Project Manager as he gave the instruction and should have notified the event to the Contractor but did not.

Various opinions have been published about the enforceability and effectiveness of time bars in contracts such as NEC3, commentators particularly debating whether the clause is a condition precedent to the Contractor being able to recover time and money, and whether a party, the Employer, can benefit from its own breach of contract to the detriment of the injured party, the Contractor.

For example, if the Employer does not provide something which he is to provide by the date for providing it shown on the Accepted Programme, this is a valid compensation event under Clause 60.1(3), but can the Employer prevent the Contractor from receiving any remedy because the Contractor failed to notify the Project Manager within the eight-week limit, and possibly, if Completion is delayed the Contractor have to pay delay damages?

The author, whilst not being a lawyer, is of the opinion that the parties are clear as to what the terms of their agreement are at the time the contract is formed. It is also clear what happens if the Contractor does not notify a compensation event within the time stated within Clause 61.3; he is protected against notifications which the Project Manager should have given but did not, and therefore the time bar must be effective and enforceable.

Clause 61.4 covers three possible outcomes to the Contractor's notification of a compensation event under Clause 61.3.

(i) Negative reply

If the Project Manager responds by stating that an event notified by the Contractor

- arises from a fault of the Contractor,
- has not happened and is not expected to happen,
- has no effect upon Defined Cost, Completion or meeting a Key Date, or
- is not one of the compensation events stated in the contract,

he notifies the Contractor of his decision that the Prices, Completion Date and the Key Dates are not to be changed. The Project Manager only needs to name one of these as his reason for refusing a compensation event.

(ii) Positive reply

If the Project Manager decides otherwise, he notifies the Contractor that it is a compensation event, and instructs him to submit quotations.

(iii) No reply

If the Project Manager does not reply within one week of the Contractor's notification, or a longer period to which the Contractor has agreed, then the Contractor may notify the Project Manager to that effect. If the Project Manager does not reply within two weeks of the Contractor's notification, then it is treated as acceptance that the event is a compensation event and an instruction to submit quotations. It is a condition precedent upon the Contractor that the notification be given before deemed acceptance can occur.

6.8 The Project Manager's assumptions

When the Contractor is instructed to submit a quotation, there may be a part of the quotation which is too uncertain to be forecast reasonably. In this case the Project Manager should state what the Contractor should assume, which could be a cost and/or time effect. Subsequently, when the effects are known or it is possible to forecast reasonably, he notifies a correction to the assumption and it is dealt with as a correction under Clause 60.1(17) (see Example under Clause 60.1(17) above). Note that the Project Manager must state the assumption; if the Contractor makes assumptions when pricing the compensation event, then they are not corrected.

Example

The Contractor is instructed by the Project Manager to excavate a trial pit to establish the nature of the substrata in a part of the Site, so that the design for a new foundation for an oil storage tank can be finalised. The Project Manager tells the Contractor that he and the Structural Engineer will inspect the sides to the excavation as the Contractor progresses with the work and will instruct the Contractor when to stop excavating.

As the plan size and depth of the excavation are uncertain, the Contractor is unable to make a reasonable forecast of proposed changes to the Prices and any delay to Completion, so the Project Manager states that the Contractor should allow an excavation 4 metres × 4 metres and to a depth of 3 metres, and to backfill the excavation on Completion when instructed. The Contractor submits his quotation which is accepted by the Project Manager, and the compensation event is then implemented.

Subsequently, the excavation is actually required to be 4 metres × 4 metres and to a depth of 4.5 metres. The Project Manager therefore notifies a correction to an assumption, which is a compensation event under Clause 60.1(17), and instructs the Contractor to submit a quotation. The resulting quotation forecasts the Defined Cost of the original trial pit and the Defined Cost of the actual trial pit and the Fee percentage is then added to the difference.

6.9 Alternative quotations

Under Clause 62.1, the Project Manager may require the Contractor to provide alternative quotations, for example to carry out the work using existing resources or, alternatively, provide additional resources which might mitigate delay. Discussion will normally take place between the Project Manager and the Contractor to ascertain what is feasible and practicable.

6.10 Delays to Completion

Note that under Clause 62.2 within the Contractor's quotation he needs to consider and include any delay to the Completion Date. Note that the clause states 'any delay', not 'any effect', so the Completion Date can only be changed to a later date or remain unchanged due to a compensation event. Note that in assessing any delay to the Completion Date the delay is the length of time that, due to the compensation event, planned Completion is later than planned Completion as shown on the Accepted Programme, so any terminal float attached to the whole programme is retained by the Contractor. Similarly, any delay to a Key Date is the length of time that, due to the compensation event, planned Completion when the Condition will be met is later than the planned Completion as shown on the Accepted Programme.

From these provisions one can see that no compensation event can result in an earlier Completion Date, though it may give rise to an earlier planned Completion. Acceleration in the ECC (Clause 36) means achieving Completion before the Completion Date which in turn, assuming the Contractor's quotation is accepted, will lead to an earlier Completion Date.

It is worth noting that the Contractor will usually show as little float as possible in his Accepted Programme, ie all operations are close to or on the critical path, so that any compensation event delays Completion and in turn the Completion Date. This has been highlighted in Chapter 3 as one of the reasons it is important for the Project Manager to spend time gaining a full understanding of the Contractor's programme and properly evaluating it before accepting it.

Once the Project Manager has instructed the Contractor to submit a quotation, the Contractor has a maximum of three weeks in which to submit his quotation to the Project Manager, although the period may be extended by agreement between the Project Manager and the Contractor under Clause 62.5.

If the Project Manager and Contractor do not agree to extend the time, once the three weeks have expired the Project Manager notifies the Contractor that he will be making his own assessment (Clause 62.3). If the Contractor has not submitted a quotation, the Project Manager notifies within two weeks of the date the Contractor should have submitted it. The Project Manager only has to notify the Contractor within the two-week period that he will be making his own assessment; the details of the assessment must be notified to the Contractor within the time period the Contractor had to submit the quotation, ie normally within a further three weeks.

Once the Contractor has submitted his quotation to the Project Manager, the Project Manager is required to reply as follows:

- He could instruct the Contractor to submit a revised quotation, in which case he explains his reasons for doing so to the Contractor – the Contractor then has a further three weeks to re-submit it, though in most cases he will review and re-submit within a shorter time scale.
- He could accept the quotation, in which case it is implemented.
- He could decide not to proceed with a proposed instruction.

Submission and assessment of quotations

Contractor submits a quotation	Project Manager replies to quotation

3 weeks	2 weeks
Time for the Contractor to submit a quotation may be extended by agreement with the Project Manager (Clause 62.5)	Time for the Project Manager may be extended by agreement with the Contractor (Clause 62.5)
If the Project Manager does not agree to extend the time for submission, he notifies the Contractor that he will be making his own assessment. (Clause 62.3)	If the Project Manager does not reply within 2 weeks the Contractor may notify the Project Manager to that effect. If the Project Manager does not reply to the notification within 2 weeks, the Contractor's notification is treated as acceptance by the Project Manager. (Clause 62.6)

Figure 6.3 Submission and assessment of quotations

- He could notify the Contractor that he will be making his own assessment, in which case the Project Manager has up to three weeks to make that assessment from the time of the notification.

If the Project Manager makes an assessment and the Contractor disagrees, then the Contractor can refer the matter to Adjudication. There is no contractual provision in NEC3 to 'agree' the Prices.

6.11 No reply to a quotation

There has always been concern amongst Contractors who were working under the first and second editions of the ECC who queried what remedy was available if the Project Manager failed to reply to a quotation within the period required by the contract. The simple answer is that failure to reply to a communication within the period required by the contract was another compensation event under Clause 60.1(6), which did not remedy the problem with the initial compensation event.

The drafters of NEC3 considered the problem and introduced a new Clause 62.6, which, in seeking to provide a remedy to the Contractor for the Project Manager's failure, seems to favour the Project Manager.

If the Project Manager does not reply to a quotation within the time allowed, the Contractor may notify him to that effect. If the Contractor has submitted more than one quotation, he states in that notification which one he proposes is to be accepted. If the Project Manager does not reply within two weeks of the notification, the notification from the Contractor is treated as acceptance of the Contractor's quotation by the Project Manager (Clause 62.6). It is somewhat concerning that a pre-condition to deemed acceptance is that the Contractor must issue a notification giving the Project Manager a further but final two weeks in which to deliberate, yet if the Contractor fails to provide a quotation within the time allowed, the Project Manager's notification tells the Contractor that he will be making his own assessment, so there is no further time for the Contractor to respond.

In addition, Secondary Option W1 (Dispute Resolution) specifically allows the Employer to refer to the Adjudicator a dispute about a quotation for a compensation event which is treated as having been accepted, so the Contractor still does not have final acceptance! This is a curious inclusion when Clause 62.6 states that no reply to the Contractor's notification is treated as acceptance by the Project Manager. One would assume that an Adjudicator whose role is to enforce the contract would have to find in favour of the Contractor.

6.12 Assessment of compensation events

The changes to the Prices are assessed as the effect of the compensation event upon:

- the actual Defined Cost of work already done;
- the forecast Defined Cost of the work yet to be done; and
- the resulting Fee.

Assessment of compensation events is based on their effect on Defined Cost and time. This is different from most standard forms of contract where variations are valued using the rates and Prices in the contract as a basis. The reason for this policy within the ECC is that no compensation event which is the subject of a quotation is due to the fault of the Contractor or relates to a matter which is at his risk under the contract. It is therefore appropriate to reimburse the Contractor his forecast additional costs or additional Defined Costs arising from the compensation event.

Under Options A to D, if the Project Manager and Contractor agree, rates and lump sums may be used instead of Defined Cost. Such rates and Prices do not have to be from the activity schedule (Options A and C) or bill of quantities (Options B and D), thereby allowing a mutually agreed fair rate or Price to be agreed.

The Contractor must ensure that he includes within his quotation for cost and time which have a significant chance of occurring and are his risk (Clause 63.6). This should include for adverse weather conditions and physical conditions which would not be compensation events.

Under Options A and C, changes to the Prices for compensation events take the form of changes to the activity schedule (Clause 63.12).

Under Clause 63.3, a delay to the Completion Date is assessed as the length of time that, due to the compensation event, planned Completion is later than planned Completion as shown on the Accepted Programme. A delay to a Key Date is assessed as the length of time that, due to the compensation event, the planned date when the Condition stated for a Key Date will be met is later than that shown on the Accepted Programme.

Clearly, this emphasises the need for an Accepted Programme to be in place in order that it can be used to make correct assessments.

No compensation event can result in a reduction in the time for carrying out the Works, ie an earlier Completion Date. Only acceleration as agreed under Clause 36 can result in an earlier Completion Date.

Any time risk allowances which the Contractor has allowed are preserved by this clause, as assessment of the compensation event is based on entitlement rather than need. Allowances for risk must be included in forecasts of Defined Cost and Completion in the same way that the Contractor allows for risks when pricing his tender. Float within the Accepted Programme is, however, available to mitigate or avoid any consequential delay to the Completion Date.

Example

On an Option C contract, the Project Manager gives an instruction to the Contractor as follows:

> Construct an additional 50 metres long × 150 mm diameter drainage run as shown on Drawing No. 32.

The Contractor's quotation will comprise changes to the Prices and also any delay to the Completion Date and any Key Dates in the contract.

Delay to Completion

First, the Contractor considers whether there will be any delay to planned Completion. The Contractor should provide details to demonstrate to the Project Manager the basis on which the delay has been calculated; this will probably require him to issue a programme, or an extract from the programme with his quotation.

He assesses that this additional work will take him two weeks to complete and as the instruction has been issued only three weeks before planned Completion, by the time he has obtained the Materials to carry out the work it will delay Completion by one week.

Delay to completion has then been assessed as one week.

Changes to the Prices

First, it must be remembered that the change to the Prices is not assessed by using the Contractor's rates from the activity schedule. It must be assessed as the effect of the compensation event upon the Defined Cost of the work and the resulting Fee.

Note that under Clause 63.15, if the Project Manager and Contractor agree, rates and lump sums may be used to assess a compensation event instead of Defined Cost. This may be done either by using rates and lump sums from the activity schedule or by simply mutually agreeing other rates and lump sums.

As this is an Option C contract, Defined Cost is assessed by using the Schedule of Cost Components.

COMPENSATION EVENT QUOTATION EXAMPLE

PEOPLE

Labourers
2 labourers × 2 weeks × £320 per week   1280.00

Ganger (part time)
1 ganger × 2 weeks × 20% × £400 per week   160.00

JCB Operator
*Note that the JCB driver is priced
with the 'people' not the 'Equipment'.*
JCB Driver
1 week × £480 per week   480.00
*As Completion is delayed, the Contractor must also
consider people costs in respect of the extended time
within the Working Areas.*

Supervision
Site Manager
1 week × £650 per week   650.00
Section Foreman
1 week × £500 per week   500.00

Total People Cost: 3070.00

EQUIPMENT
Excavator × 1 week × £1050 per week   1050.00
Fuel – 300 litres per week × 1.40 per litre   420.00
(11 litres per hour when working)
*Again, as Completion is delayed, the Contractor must
consider Equipment costs in respect of the extended
time within the Working Areas.*
Temporary Office Hire
3 × 1 week × £150 per week   450.00
Temporary Toilet Hire
2 × 1 week × £170 per week   340.00

Total Equipment Cost: 2260.00

PLANT AND MATERIALS
55 metres × 150mm diameter pipe × 14.20 per metre   781.00
(50 metres + 10% waste)

Total Plant and Materials Cost: 781.00

Charges

Item 44
Overhead costs are calculated using the formula:
Percentage for Working Areas overheads × people cost
Assume percentage for Working Areas overheads = 8%
8% × £3070.00 **Total Charges:**   245.60
                                                               6356.60
Fee percentage 10%   635.66

Total of Quotation: 6992.26

Figure 6.4 Compensation event quotation example

6.13 Changes to the Works Information that omit work

When the Project Manager gives an instruction changing the Works Information by omitting work, the method of changing the Prices and dealing with the Completion Date is often misunderstood.

The tendency is for parties to simply delete or even to just ignore the item in the pricing document (activity schedule or bill of quantities), for example in an Option A contract, the activity in the activity schedule which represents the omitted work is simply reduced or not paid, or in an Option B contract the quantity in the bill of quantities is reduced, or if a single item in the bill, the item is just not paid. Admittedly, this is usually the correct way to price change in a non-NEC contract, but in an NEC contract, both assumptions are incorrect.

Omitted work due to a change to the Works Information is assessed not by omitting or remeasuring the item in the relevant pricing document, but by forecasting the Defined Cost of that omitted work and adding the Fee percentage. The resulting amount is then adjusted against the value in the pricing document.

This principle is clearly stated in Clause 63.1 of the contract, ie the changes to the Prices are based on:

- the actual Defined Cost of work already done;
- the forecast Defined Cost of the work yet to be done; and
- the resulting Fee.

It must be remembered that the principle with compensation events in terms of their financial value is that neither party gains or loses as a result of the compensation event; the Contractor is compensated so that he is in the same financial position after the event as he would have been before the event.

If a Contractor has included low rates in his tender for work which is subsequently omitted, it is quite probable that the application of forecast Defined Cost plus Fee will give rise to a negative value; this is testament to the fact that the Contractor would have made a loss if the work had not been omitted. Similarly if high rates exist, the Contractor will retain the margin he would have made if he had carried out the work.

Example

In an Option A contract, the activity schedule includes an activity entitled 'boundary fence', the activity having a Price of £26,000. The Project Manager gives an instruction to omit the boundary fence; this is a change to the Works Information and therefore a compensation event under Clause 60.1(1).

The tendency is to assume that the Contractor is simply not paid for that activity as it will not be carried out, however the correct way to assess the change to the Prices is to forecast the Defined Cost of the work not yet done and add the resulting Fee.

Let us assume that the Contractor has already placed an order in the sum of £20,100 with a fencing Subcontractor, and can therefore prove what his costs would have been had the work not been omitted and the Fee percentage is 8%.

The forecast Defined Cost is then the value of the subcontract order:

Subcontract Order to supply and install fencing £20,100.00
Add 8%   £1,608.00

Change to the Prices   £21,708.00

The amount of the quotation, assuming it has been accepted by the Project Manager, is then used to change the Prices, so either a new activity is then inserted to the value of –£21,708 or the activity priced at £26,000 is omitted and replaced with an activity priced at £26,000 – £21,708 = £4,292.

The Contractor in this case retains the profit he would have made if the boundary fence had not been omitted. Clearly, if the change to the Prices had been more than £26,000 the Contractor would retain the loss.

The same principle would apply with all the Main Options; for example if it was an Option B contract and the boundary fence was an item in the bill of quantities, then the bill of quantities would again be changed in the same way, and again the Contractor retains the profit (or loss) he would have made if the boundary fence had not been omitted.

6.14 The Project Manager's assessments

Under Clauses 64.1–4, the Project Manager may, for the following reasons, assess a compensation event:

- if the Contractor has not submitted a required quotation and details of his assessment within the time allowed;
- if the Project Manager decides that the Contractor has not assessed the compensation event correctly in a quotation and he does not instruct the Contractor to submit a revised quotation;
- if, when the Contractor submits quotations for a compensation event, the Contractor has not submitted a programme which the contract requires him to submit; or
- if, when the Contractor submits quotations for a compensation event, the Project Manager has not accepted the Contractor's latest programme for one of the reasons stated in the contract.

These are all derived from some failure of the Contractor either to submit a quotation, to assess the compensation event correctly or to submit an acceptable programme.

If the Project Manager makes his own assessment, he should put himself in the position of the Contractor, giving a properly reasoned assessment of the effect of the compensation event, detailing the basis of his calculations and providing the Contractor with details of that assessment. A Project Manager's assessment is not simply the quotation returned to the Contractor with 'red pen' reductions down to a figure the Project Manager is prepared to accept or to get the Employer to pay!

It is also important to recognise that the Project Manager is not sending his assessment to the Contractor for his acceptance; it is the final decision under the contract, the Contractor's only remedy being adjudication.

Under Clause 64.4, if the Project Manager does not assess a compensation event within the time allowed, the Contractor may notify him to that effect. If the Project Manager does not reply to the notification within two weeks, the Contractor's notification is treated as acceptance of the quotation by the Project Manager. The only remedy the Project Manager has, if he later finds the quotation is not acceptable, is to refer it to adjudication. However, it must be remembered that the role of the Adjudicator and the adjudication process is to enforce the contract; therefore if the Contractor's quotation has been submitted in accordance with the contract, then the Employer will probably be unsuccessful in the adjudication.

6.15 Implementing compensation events

Under Clauses 65.1 and 65.2, the Project Manager implements each compensation event by notifying the Contractor of the quotation which he has accepted or of his own assessment. He implements the compensation event when he accepts a quotation or completes his own assessment or when the compensation event occurs, whichever is the latest.

These clauses emphasise the finality of the assessment of compensation events. If the subsequent records of resources on work actually carried out show that achieved Defined Cost and timing are different from the forecasts included in the accepted quotation or in the Project Manager's assessment, the assessment is not changed. The only circumstances in which a review is possible are those stated in Clause 61.6.

6.16 Final Certificate

It is worth mentioning that the ECC does not have the equivalent of a Final Account or Final Certificate which is found in other contracts, certifying that the contract has fully and finally been complied with, and that issues such as payments and compensation events have all been dealt with.

Some may say that, because of the compensation event provisions, there is no Final Account, and because the Defects Certificate confirms correction of any outstanding Defects, there is no need for a Final Certificate, however some questions could still remain such as when, under Option E, is it too late for the Contractor to submit a cost?

Appendix to Chapter 6

Suggested templates for notifying and managing compensation events.

Contract:	COMPENSATION EVENT NOTIFICATION
Contract No:	CEN No..
To: Contractor/Project Manager	

You are notified of the following compensation event:

Description

The compensation event is likely to have an effect on:

□ **Price**
□ **Completion**

Has an early warning been given?

□ **Yes**
□ **No**

Quotation required?

□ **Yes**
□ **No**

Signed: Project Manager/Contractor):............................   **Date:** ....

Copied to:

Contractor □   Project Manager □   Supervisor □   File □   Other .... □

Figure 6.5 Sample compensation event notification

Contract:	Compensation event Quotation
Contract No:	CEN No...
To: Project Manager	

Summary of Quotation

Change to Prices:

Change to Completion Date:

Signed: .... (Contractor)   Date: ....

The Project Manager's reply is:

☐ An Instruction to submit a revised quotation
☐ Acceptance of the quotation
☐ A notification that a proposed instruction will not be given
☐ A notification that the Project Manager will be making his own assessment

Signed: .... (Project Manager)   Date: ....

Copied to:

Contractor ☐   Project Manager ☐   Supervisor ☐   File ☐   Other .... ☐

Figure 6.6 Sample compensation event quotation form

Contract:

....

Contract No:

....

REGISTER OF EARLY WARNINGS AND COMPENSATION EVENTS

Ref.	Date notified	Origin PM/Sup /Cont	Description of event	Clauses	Ref.	Action by PM/C	Date Concl.	Notes
1								
2								
3								
4								
5								
6								
7								
8								
9								
10								
11								
12								
13								
14								

Figure 6.7 Sample register of early warnings and compensation events

7 Title

7.1 Introduction

Title, in the sense of ownership, is covered within the ECC by Clauses 70.1 to 73.2.

7.2 Title to Plant and Materials off Site

The general principle with title to Plant and Materials brought onto a construction Site by the Contractor is that, in the absence of any contractual provision to the contrary, title will remain with the Contractor until they are finally incorporated (built) into the Works, at which point they will become the property of the Employer. Once title has passed, the Contractor cannot, without an express right in the contract, normally with the consent of the Contract Administrator, remove the Materials.

Within the ECC, title to Plant and Materials outside the Working Areas (off Site) is covered by Clauses 70.1 and 71.1.

Clause 70.1 states that 'whatever title to Contractor has to Plant and Materials which is outside the Working Areas passes to the Employer if the Supervisor has marked it as for this contract'. The term 'marking' refers to the Supervisor either physically marking the Plant and Materials with some form of label or sign, or simply recognising that it exists by taking photographs and/or making an inventory of the Plant and Materials. This also ties in with Clause 71.1 regarding payment for Plant and Materials outside the Working Areas.

Note that in saying that 'whatever title' the Contractor has, if the Contractor does not have title himself, normally because he has not paid for it, then he does not have the authority to pass title to the Employer.

Title to Plant and Materials within the Working Areas (on Site) is covered by Clause 70.2 which states that 'whatever title the Contractor has to Plant and Materials which is outside the Working Areas passes to the Employer if it has been brought within the Working Areas'. Again, in using the words 'whatever title the Contractor has', he cannot pass a title which he does not hold himself.

Previous editions of the ECC included 'Equipment' within this clause which meant that if the Contractor owned the Equipment, title passed to the Employer when it was brought on Site, which could have caused problems! The inclusion of 'Equipment' was removed in NEC3.

Under Clause 70.2, if the Contractor removes Plant and Materials from the Working Areas with the Project Manager's permission, title reverts back to the Contractor.

Two concerns immediately come to mind:

1. One could take the requirement for permission to its extreme in that the Project Manager has to give permission for the Contractor to remove each item of redundant or waste material from the Site!
2. A Subcontractor or Supplier would not be a party to the contract, so he would not need the Project Manager's permission to remove Materials; however if he was appointed on the Engineering and Construction Subcontract he would need the Contractor's permission.

Title will then pass to the Employer once the Plant/Materials are incorporated into the Works.

Marking is also carried out if the contract identifies them for payment, and the Contractor has prepared them for marking as the Works Information requires (Clause 71.1).

When writing the Works Information and considering payment for Materials off Site, one should consider including the following to be confirmed in writing by the Contractor:

• The Plant/Materials are intended to be incorporated into the Works.
• The Plant/Materials are in accordance with the contract.
• Nothing remains to be done to the Plant/Materials prior to being incorporated into the Works, for example a steel staircase should look like a steel staircase, not a pile of unidentifiable stock steel.
• The Plant/Materials have been set aside from other stock, and clearly labelled as destined for the specific project to which the contract relates.
• The Plant/Materials are vested in the Contractor, who then transfers title in accordance with the contract, free from any charges or encumbrances.
• The Plant/Materials are insured against loss or damage for their full value under a policy of insurance protecting the interests of the Employer and the Contractor until they are delivered to the Site.
• The Plant/Materials can be inspected at the storage location at any time.
• The Plant/Materials cannot be removed from the storage location, except for delivery to Site.

The issue of title to Plant/Materials does not normally tend to cause problems unless a party becomes insolvent, when it then often becomes a major problem as a party such as an Employer who believed he had good title to unfixed Plant/Materials because the contract stated that when they were on Site so he had title, finds that he does not have title.

Whilst this book is intended for international use rather than purely focusing on English law and case law, it is not unknown for an Employer to pay a Contractor for unfixed Materials on Site, but due to the insolvency of a Subcontractor and the Contractor not having title to those Materials, the Employer then also had to pay the Subcontractor!

One aspect to bear in mind in considering the words 'whatever title the Contractor has to Plant and Materials' is the use of 'retention of title' clauses.

A retention of title clause, which normally states 'title of the goods remains with the seller until the purchase price is paid in full', is one that allows the Supplier of Materials to retain ownership of those Materials until they are fully and finally paid for irrespective of any contractual provisions or agreements between, say, the Contractor and the Employer. Sometimes a retention of title clause may state 'title in the goods remains with the seller until the price and all other sums owing by the purchaser to the seller are paid in full'.

Such clauses will override any clauses in the main contract that may stipulate that title will pass to the Employer when Materials are delivered to the Site, when they are marked, or when the Employer has paid for them as the Supplier is not party to the contract. These clauses then provide security for the Supplier against the purchaser's insolvency.

In order to be effective, retention of title clauses must be stipulated and accepted at the time the contract is formed, ie when the supplier's quotation is offered and accepted.

In order to provide this protection when insolvency occurs, the receiver or liquidator must be satisfied on each of the following points:

- The wording of the clause includes the specific goods for which protection is being claimed.
- The clause has been incorporated into the contract for the supply of those Materials.
- Any goods claimed can conclusively be identified.
- The goods supplied relate to an unpaid invoice.

The goods being claimed under the retention of title clause must be identifiable and the receiver must be fully satisfied that any items claimed are the actual goods supplied by the Supplier claiming them. Most retention of title clauses relate to raw materials, stock in trade or livestock.

The best method of identification is where the goods are marked with the name of the Supplier or where their serial numbers are quoted on any unpaid invoices. The Supplier should be allowed access to the insolvent's premises to inspect the goods which he considers are subject to his claim. Of course, the Supplier should not be allowed to remove any goods until the official receiver is satisfied that the claim is valid.

If a retention of title clause was drafted only to retain ownership of the goods until payment was made for them, goods such as any raw materials cease to be caught by it once the manufacturing process has begun, ie once the goods have lost their identity. For example, leather supplied in the production of handbags, once cut, is deemed to have created a new product.

It may be possible for the Supplier to retain title to the goods supplied even if they have been used in a manufacturing process provided they are still identifiable, in their original form and are easily removable. For instance, ceiling tiles supplied to be installed into a suspended ceiling do not lose their identity because they are fixed within the ceiling grid. However, they must be capable of being removed without damaging the ceiling grid.

7.3 Objects of value

In the event that the Contractor discovers an object of value or of other interest, such as archaeological remains within the Site, he has no title to it. He notifies the Project Manager, who then instructs the Contractor how to deal with it. It is vital that the Contractor only takes instructions from the Project Manager in this respect, although particularly in the case of archaeological remains, other parties such as local authorities, museums and educational establishments may have an interest in the find and how it is dealt with.

This is a compensation event under Clause 60.1(7) so it is important that the Contractor keeps detailed records of any stoppages, disruptions and changes to working methods as a result of the find.

7.4 Title to Materials from excavation or demolition

Unless otherwise stated in the Works Information, the Employer has the title to Materials from excavation or demolition (Clause 73.2).

It is important in this respect that the Works Information clearly details any title the Contractor has to Materials as he may wish to sell, recycle or remove those Materials, and also if the Employer has title to the Materials, he must clearly state what the Contractor is to do with those Materials, for example hand them to the Employer on Site, transport them to another location, etc.

8 Risks and insurance

8.1 Introduction

Risks and insurance, including their definition, cover requirements and policies are covered within the ECC by Clauses 80.1 to 87.3.

Insurance is, in simple terms, the principle of many paying in for the few to be compensated. It is basically a contract between the 'insured' who offers to pay an agreed sum called a premium to the 'insurer' who warrants to pay out an agreed sum should a particular and specified event occur, eg death, damage to property, etc. In effect, it is possible to insure against anything which holds an element of risk.

8.2 Types of insurance

The required insurances under a construction contract can normally be grouped under the following headings:

(i) Loss or damage to the contract Works

This covers loss or damage to the construction work itself and any unfixed Plant or Materials on Site yet to be incorporated into the Works.

(ii) Public liability

This is to cover the insured against any death, injury or damage claims from members of the public other than their own employees.

(iii) Employer's liability

In many countries, employers of staff have a statutory obligation to provide cover for their employees in the event of death, injury or damage caused during the course of their employment.

(iv) Professional Indemnity insurance

This covers parties' liabilities particularly in respect of design and the giving of advice.

8.3 Insurance requirements within the ECC

Clause 80.1 lists the Employer's risks, however, these are the Employer's risks only in terms of loss, wear or damage, and are not an exhaustive list of all the risks the Employer bears under the contract. There are many other risks such as actions or inactions of the Project Manager or Supervisor, unforeseen physical conditions and weather, which are covered elsewhere as compensation events.

There are six main categories of Employer's risk within the contract:

1. risks relating to the Employer's use or occupation of the Site, negligence, breach of statutory duty or interference with any legal right, or a fault in his design. In respect of design, the Employer should cover the risk himself through Professional Indemnity insurance, or if he uses external consultants for the design, he should ensure that they hold such insurance. This insurance should be held for the full period of liability, which will normally be several years after Completion of the Works;
2. risks relating to items supplied by the Employer to the Contractor until the Contractor has received them, or up to the point of handover to the Contractor. The Employer should ensure that he has adequate insurance in this respect, or again ensure that Others who supply items have such insurance;
3. risks relating to loss or damage to the Works, Plant and Materials, caused by matters outside the control of the parties;
4. risks arising once the Employer has taken over completed work, except a Defect that existed at take-over, an event which was not an Employer's risk, or due to the activities of the Contractor on Site after take-over;
5. risks relating to loss or wear or damage to parts of the Works taken over by the Employer, and any Equipment, Plant and Materials retained on Site after termination;
6. any other risks referred to in the Contract Data. These should be stated in Contract Data Part 1.

Any risks not carried by the Employer are carried by the Contractor from the starting date to the issue of the Defects Certificate or a Termination Certificate.

The Contractor is also obliged until the Defects Certificate has been issued and unless instructed by the Project Manager to promptly replace loss of and repair any damage to the Works, Plant or Materials. Each party indemnifies the other against any claims or proceedings which are at their risk, although if a party partly contributed to the event the other party's liability is reduced in proportion to the extent that the party contributed to that event.

The Insurance Table in the contract itemises the insurances that the Contractor has to effect together with the minimum amount of cover or minimum limit of indemnity. The default is that the Contractor provides, maintains and pays for the insurances, unless the Employer states otherwise in Contract Data Part 1.

Whether the Contractor or the Employer takes out the insurances, they are always effected as a Joint Names Policy effective until the Defects Certificate or a Termination Certificate has been issued.

8.4 The Insurance Table

The four insurances listed within the Insurance Table are as follows.

(i) Loss or damage to the Works, Plant or Materials

This item in the Insurance Table covers loss or damage to the Works, including any Plant or Materials provided by the Employer.

Some other contracts provide an option for either the Contractor or the Employer to provide this insurance, particularly where the Works relate to an existing structure or building, where the Employer may already have an existing insurance policy.

These insurances are effected as a Joint Names Policy effective until the Defects Certificate or a Termination Certificate is issued and includes full reinstatement, replacement or repair (the amount stated should also include an amount or percentage to cover Professional Fees).

Insurance of the Works can often be covered by the Contractors' All Risks (CAR) policy. The minimum cover for all insurances is stated in the Insurance Table, however the Contractor is liable for whatever the amount of any claim therefore he must consider the minimum value in the Insurance Table purely as a guide.

(ii) Loss or damage to Equipment

Again, the CAR policy should cover this. The reference to 'replacement cost' means the cost of replacement with Equipment of similar age and condition rather than 'new for old'.

(iii) Loss or damage to property (except the Works, Plant or Materials) or bodily injury or death of a person not an employee of the Contractor

This requires the Contractor to indemnify the Employer against any loss, expense, claim, etc in respect of any personal injury or death caused by the carrying out of the work, other than their own employees. This includes the liability towards members of the public who may be affected by the construction work, although they have no part in it. In the case where a party makes a claim directly against the Employer due to a death or injury, the Contractor should either take on that claim, or alternatively the Employer can sue the Contractor to recover any monies.

The Insurance Table states the minimum amount of cover or minimum limit of indemnity. The Contractor could be liable for whatever the amount of any Claim, therefore he must consider the minimum value in the Insurance Table purely as a guide.

The Contractor must be able to prove, if requested by the Employer, that he has the required cover and his premiums are up to date, as if he does not, the Employer may take out insurances himself and deduct the premiums from the

Contractor. The Contractor also used to have to prove that his Subcontractors had the same cover, however this has now been amended as Contractors can either ensure that the Subcontractors have the cover, or can include it under their own cover on behalf of their Subcontractors.

(iv) Death of or bodily injury to employees of the Contractor

This covers the liabilities as an employer of people to insure against injury or death caused to people whilst carrying out their work, which is a legal obligation in most countries. The Contractor must be able to prove, if requested by the Employer, that he has the required cover and his premiums are up to date, as if he does not, the Employer may take out insurances himself and deduct the premiums from the Contractor. The Contractor also used to have to prove that his Subcontractors had the same cover; however, this has now been amended as Contractors can either ensure that the Subcontractors have the cover, or can include it under their own cover on behalf of their Subcontractors.

Again, the Insurance Table states the minimum amount of cover or minimum limit of indemnity, the Contractor having to insure for whatever the amount of any claim.

8.5 Insurance policies

The Contractor is required to submit to the Project Manager for acceptance certificates of insurance as required by the contract signed by the insurer or the insurance broker:

- before the starting date;
- on each renewal of the insurance policy.

If the Contractor does not submit a required certificate, the Employer may insure a risk, the cost of the premium being paid by the Contractor.

If the Employer is to provide insurance, then the Project Manager submits policies and certificates to the Contractor for acceptance before the starting date and as and when instructed by the Contractor.

If the Employer does not submit a required certificate, the Contractor may insure a risk, the cost of the premium being paid by the Employer.

8.6 Contractors' All Risks Insurance

Contractors' All Risks (CAR) Insurance normally covers damage to property, such as damage to buildings and other structures being constructed or to the existing building in which the construction is being carried out. It also normally covers liability for third party claims for injury and death or damage to third party property.

The basic principle is that CAR covers those losses not covered by an 'Excluded Risk'. For example, the JCT contracts exclude from cover risks such as Defects due to 'wear and tear, obsolescence, deterioration, rust and mildew, loss and damage arising out of war and for faulty workmanship and faulty design'. The benefit to the insured under this type of policy is that the burden is shifted to the insurer who is required to show that the cause of the loss falls within an exclusion.

This insurance is usually taken out in the joint names of the Contractor and the Employer. Other interested parties, such as funders, often ask to be added as a joint name. The theory is that if damage occurs to the insured property then, regardless of fault, insurance funds will be available to allow for reinstatement.

The effect of Joint Names Insurance is that each party has its own rights under the policy and can therefore claim against the insurer. Each insured should comply with the duties of disclosure and notification.

8.7 Insurance of neighbouring property

Some contracts have provision for insuring neighbouring property which may suffer from collapse, subsidence, vibration, etc due to damage by, for example, piling, deep excavations or demolition works, where not caused by the negligence of the Contractor.

The ECC does not provide for this, but it can be added as a Z clause if applicable.

The Employer should consider his liabilities in respect of neighbouring owners who may be affected by the Works and, if necessary, either instruct the Contractor to provide such insurance, or take it out himself.

Such a policy would normally exclude:

- damage caused by the negligence of the Contractor or his Subcontractors;
- damage caused by errors or omissions in the Employer's design;
- damage which can be reasonably foreseen, eg cracking may have already occurred to a neighbouring building prior to the Works commencing – in this respect, it is important that a condition survey with photographic evidence be carried out before construction commences;
- where the damage is caused by Excepted Risks – these risks will normally be held by the Employer.

8.8 Professional Indemnity or Professional Liability insurance

Contract Data Part 1 provides for the Employer to insert a requirement for the Contractor to provide additional resources.

Where the Contractor is designing parts of the Works, it would be advisable for the Employer to include a requirement for Professional Indemnity (PI) insurance.

If a Contractor or Consultant providing a service to an Employer makes a mistake, is found to be negligent, or gives inaccurate advice, then he will be liable to the Employer in the event that the Employer incurs a loss as a result. This loss can be very significant where the design has to be corrected, parts of the structure have to be taken down and reinstated, a facility has to be closed down whilst the remedial measures take place, and there are also legal costs.

PI claims can arise where there is negligence, misrepresentation or inaccurate advice which does not give rise to bodily injury, property damage or personal injury, but does give rise to some financial loss.

Additional coverage for breach of warranty, intellectual property, personal injury, security and cost of contract can be added.

In that event, although the Employer claims against the Contractor or Consultant rather than from the insurers, PI insurance protects the Contractor or Consultant against claims for loss or damage made by an Employer or third party.

PI insurance policies are based on a 'claims made' basis, meaning that the policy only covers claims made during the policy period when the policy is 'live', so claims which may relate to events occurring before the coverage was active may not be covered. However, these policies may have a retroactive date which can operate to provide cover for claims made during the policy period but which relate to an incident after the retroactive date. The Project Manager, on behalf of the Employer, should ensure that the Contractor has taken out and maintained the required insurances for the full period of his liability.

Claims which may relate to incidents occurring before the policy was active may not be covered, although a policy may sometimes have a retroactive date, earlier than when the policy was created.

9 Termination

9.1 Introduction

Termination, including reasons, termination procedures and payment, are covered within the ECC by Clauses 90.1 to 93.2 with further references within the Main Option clauses.

Termination is a word most commonly used in the context of construction contracts to refer to the ending of the Contractor's employment. The parties have a common law right to bring the contract to an end in certain circumstances, but most standard forms give the parties additional and express rights to terminate upon the happening of specified events.

Some contracts refer to termination of the contract, whilst others refer to the termination of the Contractor's employment under the contract. In practice, it makes little difference, and most contracts make express provision for what is to happen after termination.

It should be noted that most contracts require a notice to be given by one party to the other unless the breach is due to insolvency, the offending party then has a period of time in which to remedy the breach, failing which termination can take place.

It is important to note that neither party should attempt to terminate employment under the contract unless they are sure that the provision is available within the contract. If termination is held to be wrongful, it is usually a repudiation of the contract.

Termination is an entitlement not an obligation.

9.2 Termination within the ECC

Within the ECC, there are 21 reasons listed for either party to terminate:

- *Reasons R1 to R10* provide for either party to terminate due to the insolvency of the other. This includes bankruptcy, appointment of receivers, winding-up orders and administration orders dependent on whether the party is an individual, a company or a partnership.
- *Reasons R11 to R13* provide for the Employer to terminate if the Project Manager has notified the Contractor that he has defaulted and the

Contractor has not remedied the default within four weeks of the notification in respect of the Contractor substantially failing to comply with his obligations, not providing a bond or guarantee, or appointing a Subcontractor for substantial work before the Project Manager has accepted the Subcontractor.

It is quite subjective as to what the words 'substantially failed to comply' and 'substantial work' means within these provisions. One would assume that it is not intended to relate to minor breaches, but the Contractor could 'substantially fail to comply' with a minor obligation.

- *Reasons R14 to R15* provide for the Employer to terminate if the Project Manager has notified that the Contractor has defaulted and the Contractor has not stopped defaulting within four weeks of the notification in respect of substantially hindering the Employer or Others or substantially broke a health or safety regulation and not stopped defaulting within the four weeks of the notification.

Again, the use of the word 'substantially' is quite subjective. How would one substantially (or unsubstantially) hinder someone, or substantially (or unsubstantially) break a health or safety regulation? Is not a health or safety regulation either 'broken' or 'not broken'?

- *Reason R16* provides for the Contractor to terminate if the Employer has not paid an amount certified by the Project Manager within 13 weeks of the date of the certificate.

- *Reason R17* provides for either party to terminate if the parties have been released under the law from further performance of the whole contract. Note the reference to 'the whole contract', not just a part.

- *Reasons R18 to R20* provide for the Project Manager having instructed the Contractor to stop or not restart any substantial work or all work and an instruction allowing the work to restart or start has not been given within 13 weeks.

Either party may terminate if the instruction was due to a default by the other, or if the instruction was due to any other reason.

- *Reason R21* provides for the Employer to terminate if an event occurs which stops the Contractor completing the Works, or stops the Contractor completing the Works by the date shown on the Accepted Programme and is forecast to delay Completion by more than 13 weeks, and which neither party could prevent and an experienced Contractor would have judged at the Contract Date to have such a small chance of occurring that it would have been unreasonable for him to have allowed for it.

Note that the Contractor can only terminate for one of the above reasons, though the Employer can terminate for any reason. Whilst this may seem inequitable, if the Employer terminates for a reason not stated in the Termination Table, he will have to pay the Contractor for work carried out up to the termination, other costs incurred in the expectation of completing the whole of the Works, for example orders placed and other commitments made, the forecast Defined Cost of removing Equipment, and dependent on the Main Option used, the direct Fee percentage applied to the difference between the original Total of the Prices and the Price for Work Done to Date, essentially a loss of profit/overheads provision. Clearly, the earlier the Employer exercises this right of termination, the greater this amount.

With all the termination reasons, the party wishing to terminate notifies the Project Manager and the other party giving details of his reasons for terminating. The Project Manager then issues a Termination Certificate to both parties if he is satisfied that the reason for termination is valid under the contract. It is perhaps curious that the Project Manager, who acts for the Employer, makes the decision as to whether the termination is valid.

9.3 Procedures on termination

Under any of the reasons, whoever terminates, and whether the reason is included in the Termination Table or not, the Employer may complete the Works, either himself or using another Contractor, and may use any Plant and Materials to which he has title.

Employer terminates

If the termination has arisen for any reason other than that stated in the Termination Table, the Employer may instruct the Contractor to leave the Site, remove any Equipment, Plant and Materials to which he has title and assign the benefit of any contract or subcontract to the Employer.

For Reasons R1 to R15 or R18, the Employer may instruct the Contractor to leave the Site and remove any Equipment, Plant or Materials which belong to the Contractor and assign the benefit of any subcontract to the Employer.

For Reasons R1 to R15, R17, R18 or R20, the Employer may use any Equipment to which the Contractor has title. The Contractor removes such Equipment when instructed to do so. Clearly in this respect, the Contractor does not have title to hired Equipment; also, if the Equipment is owned by the Contractor but he has become insolvent, then some discussion may have to take place between the Employer and the receivers or other authorised representatives.

For Reason R21 the Contractor leaves the Site and removes Equipment.

Contractor terminates

Whatever the reason, the Contractor leaves the Site and removes Equipment.

9.4 Payment on termination

Under any of the reasons, whether included in the Termination Table or not, on termination, the Contractor is entitled to be paid:

- an amount due as for normal payments;

- the Defined Cost for Plant and Materials within the Working Areas, or to which the Employer has title;
- other Defined Cost in reasonable expectation of completing the whole of the Works;
- any amounts retained by the Employer;
- a deduction of any repaid balance of an advanced payment.

Apart from when the Employer terminates for Reasons R1 to R15 or R18, the Contractor is paid the forecast cost of removing Equipment.

If the Employer terminates for Reasons R1 to R15 or R18, a deduction is made of the forecast of the additional cost to the Employer of completing the Works.

If the Employer terminates for a reason other than one stated in the Termination Table as stated above, dependent on the Main Option used, the direct Fee percentage is applied to the difference between the original Total of the Prices and the Price for Work Done to Date. If the Contractor terminates for Reason R1 to R10, R16 or R19, the same provision applies.

9.5 Termination and Subcontractors

The parties should also consider how termination of their contract affects Subcontractors. If the Contractor is the defaulter, termination occurs and the Employer wishes to proceed with the Works, any subcontracts would be assigned to the Employer.

However, if the Contractor is the terminating party he should ensure that provisions are in place regarding termination of subcontracts, and termination provisions in the main contract should be mirrored in the subcontracts.

10 Disputes

10.1 Introduction

Dispute resolution is provided for within the ECC by Secondary Options W1 and W2, which cover the adjudication process, whether the UK-based legislation, the Housing Grants, Construction and Regeneration Act 1996 is applicable (Option W2) or not applicable (Option W1). Note that the Housing Grants, Construction and Regeneration Act 1996 has now been supplemented by the Local Democracy, Economic Development and Construction Act 2009.

If the ECC is used within the UK, the Employer must by reference to the Act determine whether the contract is a 'construction contract', and select the appropriate dispute resolution option. If the ECC is used outside the UK, Option W1 will always apply.

Clearly, there is nothing to prevent the parties jointly attempting to resolve a dispute by a consensual and non-binding route such as negotiation, mediation, conciliation or mini trial prior to referring it to adjudication.

10.2 What is adjudication?

Adjudication has been defined as 'a summary non-judicial dispute resolution procedure that leads to a decision by an independent person that is, unless otherwise agreed, binding upon the parties, but which may subsequently be reviewed by means of arbitration, litigation or by agreement'.

It is therefore an intermediate dispute resolution process which is binding upon the parties unless and until the matter is referred to arbitration or litigation.

In the UK, adjudication is a statutory right within a construction contract, but it has always been included as the primary means to resolve a dispute within the NEC3 contracts, failing which the parties may refer a dispute to a tribunal. In fact, the majority of adjudication decisions are accepted by parties as the final result and are never taken to the tribunal.

10.3 Who can be an Adjudicator?

Whilst many professional bodies hold registers of Adjudicators there is no formal qualification to become an Adjudicator, though there are commonly acceptable requirements which include:

- A recognised professional qualification in a relevant discipline. The relevant discipline then normally relates to the dispute in question; for example, if the dispute is related to services installations, a Mechanical and Electrical Engineer would probably be a relevant discipline, though it must be remembered that the Adjudicator's role is to enforce the contract, so in that case he does not necessarily have to be a Mechanical and Electrical Engineer just because the dispute is related to services.
- Substantial number of years' experience within the construction industry, normally 10 to 15 years post qualification in a relevant discipline.
- A qualification, or at least a thorough understanding of contracts and legal principles, though in most cases the Adjudicator is not a lawyer, nor does he have to be. Many say it is better if he is not a lawyer, but that is probably a view left to the cynics!
- Recognised training in the law and practice of adjudication.
- Membership of a professional institution.
- Some also say that the Adjudicator should have Professional Indemnity (PI) insurance, though it must be emphasised that as the Adjudicator is not liable to the parties for any action or inaction on his part other than in bad faith, it could be questioned as to whether such insurance should be a prerequisite to appointment.

As the use of adjudication in the resolution of contractual disputes has grown, particularly in the UK where parties have a statutory right to refer a dispute to adjudication, many education establishments and professional bodies now provide diploma and certificate courses in adjudication.

10.4 Who pays for the adjudication?

The cost of the Adjudicator's fee is shared between the parties, the Adjudicator normally deciding what the proportion is to be paid by each party, with in some cases one party paying the majority or even all of the Adjudicator's fee. However, both parties are jointly and severally liable for payment of the Adjudicator's fee; therefore if one party will not or cannot pay his share, possibly due to insolvency, the other party must pay it, and recover the amount directly from the defaulting party.

Each party bears his own costs in collating, copying, postage, etc which is different to, for example, litigation where both parties' costs are usually borne by the losing party.

10.5 What will the Adjudicator charge?

There are no set hourly rates for Adjudicators; each Adjudicator will advise the party of his hourly rate and whether there are other costs payable, for example for travel or printing. The Adjudicator is required to properly record the time spent carrying out his duties.

10.6 What if a party does not comply with the Adjudicator's decision?

If a party does not comply with the Adjudicator's decision, the other party can enforce that decision by going to court.

10.7 Selecting the Adjudicator

Within ECC Contract Data Part 1 there are three options for selecting the Adjudicator:

1. The name and address of the Adjudicator may be stated by the Employer in Contract Data Part 1 and he is then appointed before the starting date. This has the advantage that tendering Contractors are aware at the time of tender who will be the Adjudicator in the event that a dispute arises and, if necessary, can object to them within their tenders. Also, the Adjudicator is already in place should a dispute arise. The named Adjudicator may require some form of retainer fee for being named in the contract and being available in the event that a dispute arises.
2. The parties can mutually agree to the name of the Adjudicator in the event that a dispute arises. This has the advantage that there is no 'Adjudicator in waiting' and therefore no fee payable. Many ECC practitioners say that pre-appointing the Adjudicator as in Option 1 is resigning oneself to the fact that there will at some point be a dispute. However, once parties are in dispute they then have to agree who is to be the Adjudicator and appoint him; at this point, though, they may not wish to agree with each other about anything!
3. The name of the Adjudicator Nominating Body may be stated by the Employer in Contract Data Part 1. This is probably the favoured option as an independent name can be put forward by the nominating body, who is normally well prepared to put forward the name of an Adjudicator and have him appointed within the time scales set by the contract and if necessary the appropriate legislation.

In all cases, the Adjudicator must be impartial, ie the Adjudicator should not show any bias towards either party. In addition, all correspondence from the Adjudicator must be circulated to both parties.

Any request for an Adjudicator must be accompanied by a copy of the notice of adjudication, and the appointment of the Adjudicator should take place within seven days of the submission of the notice of adjudication to the other party.

10.8 Adjudicator Nominating Bodies

Adjudicator Nominating Bodies, as the name suggests, are organisations that fulfil the role of nominating Adjudicators. These bodies keep registers of Adjudicators with the required expertise and based at various geographical locations who can act for parties should they be nominated.

A list of recognised Adjudicator Nominating Bodies in the UK is as follows:

- Association of Independent Construction Adjudicators (AICA);
- Centre for Effective Dispute Resolution (CEDR);
- Chartered Institute of Arbitrators (CIArb);
- Chartered Institute of Arbitrators (Scotland) (CIArb-Scotland);
- Chartered Institute of Building (CIOB);
- Construction Conciliation Group (CCG);
- Construction Confederation (CC);
- Construction Industry Council (CIC);
- Construction Plant-Hire Association (CPA);
- Dispute Board Federation (DBF);
- Dispute Resolution Board Foundation (DRBF);
- Institution of Chemical Engineers (IChemE);
- Institution of Civil Engineers (ICE);
- Institution of Electrical Engineers (IEE);
- Institution of Mechanical Engineers (IMechE);
- Law Society of Scotland (LawSoc (Scot));
- Nationwide Academy of Dispute Resolution (NADR);
- RICS Dispute Resolution Service (RICS DRS);
- Royal Incorporation of Architects in Scotland (RIAS);
- Royal Institute of British Architects (RIBA);
- Royal Institution of Chartered Surveyors (RICS);
- Royal Society of Ulster Architects (RSUA);
- Technology and Construction Court Bar Association (TECBAR);
- Technology and Construction Solicitors Association (TeCSA).

Clearly, in naming the Adjudicator Nominating Body in the contract it is advisable to name an organisation with expertise in the project to be carried out.

These bodies are very knowledgeable about appointment of Adjudicators and the relevant time scales and for a modest fee can nominate a suitably qualified Adjudicator to suit the parties' requirements.

10.9 Preparing for adjudication

Once the adjudication process starts by the issuing of a notification of dispute by the referring party it is a very quick procedure, therefore it is critical that the referring party, and in turn the responding party, prepare well in the short time they have available. Time scales are stated within the contract, though these may be extended by agreement between the parties.

As the process does not start until the referring party initiates the process by issuing the notification of the dispute to the other party, the referrer does have the advantage of time, and in many cases surprise, as he can prepare his case before issuing the notification.

Before considering initiating the adjudication process, parties should consider the following:

- There must be a dispute. Whilst this may seem fairly obvious it is essential that a dispute actually exists between the parties in that one party has made a claim, which the other party denies (probably repeatedly), and that the parties have exhausted the means to resolve the dispute themselves, for example by the use of negotiation, mediation, conciliation etc, though they do not have to prove that they have tried other methods.
- The dispute arises under the contract. The Adjudicator is appointed to decide a dispute under the contract; therefore it is essential that the dispute arises on a matter covered by the contract. If the dispute is whether a contract exists in the first place, then an Adjudicator cannot decide that.
- The role of the Adjudicator is to enforce the contract, not to decide who is the more deserving party. It is therefore essential that, before commencing adjudication, the referring party ensures that he has properly complied with the contract; otherwise he may find himself in a weak position.

  There have been adjudications in the past where a Contractor has acted in good faith but not strictly in accordance with the contract, accepting and willingly acting on verbal instructions in order not to damage what appeared to be a good relationship with the Employer, then when a dispute arises the Adjudicator states that as the contract did not provide for verbal instructions, the Contractor was incorrect to act on them, and therefore the Contractor loses the adjudication.
- Whilst many people say that, as adjudication is a fairly simple process, it should not include lawyers as the issues are construction-related rather than legal issues, it is important that the parties carry out their obligations within the short time scale they have, so if a party is uncertain as to what they have to do, and when and how they have to do it, they obtain professional advice, which may be from a lawyer.
- It is essential that a clear and concise notice of adjudication is prepared by the referring party as it defines what is in dispute, and the matters that are being referred to the Adjudicator. The Adjudicator cannot consider issues which are not set out within the notice of adjudication.

  The notice should include:

  - Brief details of the project. Whilst the Adjudicator may have been named in the contract, he may have little knowledge about the actual project.
  - The names and addresses of the parties to the contract and the parties in dispute.
  - The nature and a brief description of the dispute.
  - The nature of the redress which is sought from the Adjudicator. This is critical; many referring parties do not make it clear what they want the Adjudicator to decide!

- Details included within the referral should be clearly and concisely presented so that the responding party knows what it needs to respond to, and also the Adjudicator clearly understands what is required of him. Any supporting information should be clearly referenced and relative to the matter in dispute. Neither party should not include or rely on any information or evidence which the other party has not seen. If they do, the other party may raise an objection, which could bring the adjudication process to a premature end without a resolution.
- The contract provides for time scales to be extended by agreement between the parties, so if more time is needed in order to prepare and submit documentation, the other party should be consulted.

10.10 Statutory right to adjudication (UK-based contracts)

When the Housing Grants, Construction and Regeneration Act 1996 was introduced in the UK, it applied to construction contracts which came into existence after 1 May 1998, and apart from a few notable exceptions, it provided a statutory right to refer a dispute under a construction contract to adjudication. This Act has now been supplemented by the Local Democracy, Economic Development and Construction Act 2009.

10.11 What do the Housing Grants, Construction and Regeneration Act 1996 and the Local Democracy, Economic Development and Construction Act 2009 cover?

The Act applies to all construction contracts including:

- all normal building and civil engineering work, including construction, alteration, repair, maintenance, extension and demolition or dismantling of structures forming part of the land and works forming part of the land, whether they are permanent or not;
- the installation of mechanical, electrical and heating works and maintenance of such works;
- cleaning carried out in the course of construction, alteration, repair, extension, painting and decorating and preparatory works;
- agreements with consultants such as Architects, other Designers, Engineers and Surveyors;
- elements such as scaffolding, Site clearance, painting and decorating;
- labour-only contracts;
- contracts of any value.

It excludes:

- work on process plant and on its supporting or access steelwork on Sites where the primary activity is nuclear processing, power generation, water or effluent treatment, handling of chemicals, pharmaceuticals, oil, gas, steel or food and drink;

- supply-only contracts, that is contracts for the manufacture or delivery to Site of goods where the contract does not provide for their installation;
- extracting natural gas, oil and minerals (and the workings for them);
- purely artistic work;
- off-site manufacture;
- contracts with residential occupiers – this means a contract where work is carried out on a dwelling which one of the parties occupies or intends to occupy as his residence rather than simply a housing project;
- PFI contracts – this only includes the head agreements in PFI projects, not a contract for the construction of the Works;
- finance agreements – this includes loan agreements.

The Adjudicator is appointed jointly by both parties who are jointly and severally liable for those fees.

Under the Act, the contract must include an adjudication clause which complies with the requirements of the Act. If the contractual clause does not comply then it is voidable and the parties can exercise their right to adjudication under the 'Scheme for Construction Contracts' to be issued by the Secretary of State. The provisions in the contract must:

- enable the parties to give notice at any time;
- provide a timetable with the object of securing the appointment of the adjudicate and referral of the dispute to him within seven working days;
- require the Adjudicator to reach a decision within 28 days of referral or such longer period as is agreed by the parties after the dispute has been referred;
- allow the Adjudicator to extend the period of 28 days by up to 14 days with the consent of the party by whom the dispute was referred;
- impose a duty on the Adjudicator to be impartial;
- enable the Adjudicator to take the initiative in ascertaining the facts and the law;
- provide that the decision of the Adjudicator is binding until determined by litigation, arbitration or by agreement;
- provide that the Adjudicator is not liable for anything done or omitted in the discharge or purported discharge of his functions as Adjudicator unless the act or omission is in bad faith, and that any employee or agent of the Adjudicator is similarly protected from liability.

10.12 Does a 'dispute' exist?

Only a dispute between the contracting parties can be referred to adjudication. If there is no dispute, then an Adjudicator has no jurisdiction to consider the matter.

As an example, on a non-NEC contract if the Contractor submits a loss and expense claim and the Contract Administrator does not agree with it, then the simple fact that the claim is not agreed does not mean that there is a dispute as the parties are still free to meet and, if necessary, to negotiate a settlement. It is only when the parties have exhausted all the opportunities to meet and reach a compromise, but still have a difference that a dispute exists, and the matter can be referred to the Adjudicator.

It must also be noted that an Adjudicator only has jurisdiction to consider one dispute at a time under the same contract unless there is agreement between the disputing parties. Therefore, if there are a number of issues in dispute, unless they are inextricably joined then there are likely to be several adjudications.

Disputes under or in connection with the contract may be referred to the Adjudicator. Note that a dispute under the contract cannot be referred to the tribunal (arbitration or litigation) unless it has first been decided by the Adjudicator.

Adjudication is a quick, convenient and relatively inexpensive way of resolving a dispute, whereby an impartial third party Adjudicator either named in the contract, selected by agreement between the parties or nominated by a third party, decides the issues in dispute between the parties. It is quicker and less expensive than arbitration or litigation, though the parties may refer to one of those methods following adjudication.

It must be recognised that the role of the Adjudicator is to enforce the contract, not to decide what, in his opinion, would be the fairest outcome. For example, if the parties have a dispute over the quality of suspended ceilings carried out, it is the Adjudicator's task to ascertain whether the ceilings were supplied and installed in accordance with the contract, ie drawings, specifications, etc and the applicable law, not whether it was a good or bad job. In addition, the question of whether a contract exists cannot be decided by the Adjudicator but more likely through the courts.

10.13 Adjudication provisions within the ECC

The ECC provides two Options for referral to the Adjudicator.

(i) Option W1

Option W1 is used unless the UK Housing Grants, Construction and Regeneration Act 1996 applies.

For this Option there is an Adjudication Table under which, if the referring party is the Contractor and the dispute is due to an action or inaction of the Project Manager or Supervisor he must notify the dispute to the Employer within four weeks of becoming aware of the action or inaction, and then refer it to the Adjudicator between two and four weeks after that notification.

If the referring party is the Employer and the dispute is regarding a compensation event which has been treated as having been accepted (Clause 62.6) then the Employer must notify the dispute to the Contractor and then refer it to the Adjudicator between two and four weeks after that notification. This seems an odd reason for referring a dispute to the Adjudicator, as Clause 62.6 states that if the Project Manager does not reply to the Contractor's notification it is treated as acceptance of the quotation, so how could an Adjudicator whose role is to enforce the contract find other than in favour of the Contractor?

Some practitioners have stated that this inclusion could in theory allow the Employer to override the actions of his own Project Manager but the referral to the Adjudicator is initiated by the Project Manager notifying the dispute to the Employer and the Contractor so the Employer could not initiate the action himself.

For any other matter which the Employer wishes to refer to the Adjudicator, it may be referred between two and four weeks after notifying the Contractor and the Project Manager.

Whilst these time scales are fixed within the Adjudication Table, they can be extended by the Project Manager if the Contractor and the Project Manager agree before either the notice or the referral is due. Note that if the matter in dispute is not notified and referred within these time scales, neither party can refer the dispute to the Adjudicator or the tribunal.

The Adjudicator is appointed under the NEC3 Adjudicator's Contract, he acts impartially, and if he resigns or is unable to act the parties jointly appoint a new Adjudicator. One would assume that the referring party obtains a copy and completes the Adjudicator's Contract. If the parties have not appointed an Adjudicator, either party may ask the nominating body to choose an Adjudicator within four days of the request.

No procedures have been specified for appointing a suitable person, and in practice a number of different methods have been used. Whatever method is used, it is important that both parties have full confidence in his impartiality, and for that reason it is preferable that a joint appointment is made. The Adjudicator should be a person with experience in the type of work included in the contract between the parties and who occupies or has occupied a senior position dealing with disputes. He should be able to understand the viewpoint of both parties.

Often the parties delay selecting an Adjudicator until a dispute has arisen, although this frequently results in a disagreement over who should be the Adjudicator. As noted, the selection of the Adjudicator is important, and it should be recognised that a failure to agree an Adjudicator means that a third party will make the selection without necessarily consulting the parties.

The referring party must include within his referral information which he wishes to be considered by the Adjudicator, any more information from either party to be provided within four weeks of the referral.

A dispute under a subcontract, which is also a dispute under the ECC, may be referred to the Adjudicator at the same time and the Adjudicator can decide the two disputes together.

The Adjudicator may review and revise any action or inaction of the Project Manager, take the initiative in ascertaining the facts and the law relating to the dispute, instruct a party to provide further information and instruct a party to take any other action which he considers necessary to reach his decision.

All communications between a party and the Adjudicator must be communicated to the other party at the same time.

The Adjudicator decides the dispute and notifies the parties and the Project Manager within four weeks of the end of the period for providing information. This four-week period may be extended by joint agreement between the parties. Until this decision has been communicated, the parties proceed as if the matter in dispute was not disputed.

The Adjudicator's decision is binding on the parties unless and until revised by a tribunal and is enforceable as a contractual obligation on the parties. The Adjudicator's decision is final and binding if neither party has notified the Adjudicator that they are dissatisfied with an Adjudicator's decision within the time stated in the contract, and intends to refer the matter to the tribunal.

The Adjudicator may, within two weeks of giving his decision to the parties, correct a clerical mistake or ambiguity.

Review by the tribunal

A dispute cannot be referred to the tribunal unless it has first been referred to the Adjudicator.

The tribunal may be named by the Employer within Contract Data Part 1. Whilst no alternatives are stated, it will normally be litigation or arbitration. If arbitration is chosen the Employer must also state in Contract Data Part 1, the procedure, the place where the arbitration is to be held, and the person or organisation who will choose an Arbitrator if the parties cannot agree a choice, or if the named procedure does not state who selects the Arbitrator.

A party can, following the adjudication, notify the other party within four weeks of the Adjudicator's decision that he is dissatisfied. This is a time-barred right, as the dissatisfied party cannot refer the dispute to the tribunal unless it is notified within four weeks of the Adjudicator's decision, failure to do so will make the Adjudicator's decision final and binding.

Also, if the Adjudicator has not notified his decision within the time provided by the contract, a party may within four weeks of when the Adjudicator should have given his decisions notify the other party that he intends to refer the dispute to the tribunal.

The tribunal settles the dispute and has the power to reconsider any decision of the Adjudicator and review and revise any action or inaction of the Project Manager or Supervisor.

It is important to note that the tribunal is not a direct appeal against the Adjudicator's decision; the parties have the opportunity to present further information or evidence that was not originally presented to the Adjudicator, and also the Adjudicator cannot be called as a witness.

(ii) Option W2

The inclusion of adjudication within the New Engineering Contract, now the NEC3, pre-dates UK legislation giving parties to a contract the statutory right to refer a dispute to adjudication.

The Local Democracy, Economic Development and Construction Act 2009

The Local Democracy, Economic Development and Construction Act 2009 which came into force in England and Wales on 1 October 2011 and in Scotland on 1 November 2011 amended the Housing Grants, Construction and Regeneration Act 1996, the new Act applying to most UK contracts after that date. There still remain certain categories of contract to which neither Act applies.

The differences between the Housing Grants, Construction and Regeneration Act 1996 and the Local Democracy, Economic Development and Construction Act 2009 have been explored in a previous chapter but in respect of adjudication they include:

- The earlier Act only applied to contracts which were in writing, whereas the new Act also applies to oral contracts.
- Terms in contracts such as 'the fees and expenses of the Adjudicator as well as the reasonable expenses of the other party shall be the responsibility of the party making the reference to the Adjudicator' have been prohibited. Much has been written about the effectiveness of such clauses, often referred to as Tolent clauses (*Bridgeway Construction Ltd v Tolent Construction Ltd 2000*), and whether they comply with the earlier Act, so the new Act should provide clarity for the future.
- The Adjudicator is permitted to correct his decision so as to remove a clerical or typographical error arising by accident or omission. Previously he could not make this correction.

Section 108 of the new Act provides parties to construction contracts with a right to refer disputes arising under the contract to adjudication. It sets out certain minimum procedural requirements which enable either party to a dispute to refer the matter to an independent party who is then required to make a decision within 28 days of the matter being referred.

If a construction contract does not comply with the requirements of the Act, or if the contract does not include an adjudication procedure, a statutory default scheme, called the Scheme for Construction Contracts (referred to as the 'Scheme'), will apply.

The Act provides that a dispute can be referred to adjudication 'at any time' provided the parties have a contract. This is not necessarily during the construction stage, for example a Designer may refer a dispute during the design stage.

'At any time' can also refer to a dispute after the contract is completed.

What are the requirements?

Section 108 of the Act requires all 'construction contracts', as defined by the Act, to include minimum procedural requirements which enable the parties to a contract to give notice at any time of an intention to refer a dispute to an Adjudicator. The contract must provide a timetable so that the Adjudicator can be appointed, and the dispute referred, within seven days of the notice. The Adjudicator is required to reach a decision within 28 days of the referral, plus any agreed extension, and must act impartially. In reaching a decision an Adjudicator has wide powers to take the initiative to ascertain the facts and law related to the dispute.

For this Option there is no Adjudication Table as the parties have a right to refer a dispute to each other and to the Adjudicator at any time.

The Housing Grants, Construction and Regeneration Act 1996 defines adjudication as:

> a summary non-judicial dispute resolution procedure that leads to a decision by an independent person that is, unless otherwise agreed, binding upon the parties for the duration of the contract, but which may subsequently be reviewed by means of arbitration, litigation or by agreement.

In that sense, adjudication does not necessarily achieve final settlement of a dispute because either of the parties has the right to have the same dispute heard afresh in court, or, where the contract specifies, arbitration.

However, experience since the Housing Grants, Construction and Regeneration Act 1996 came into force shows that the majority of adjudication decisions are accepted by the parties as the final result.

The Adjudicator is appointed under the NEC Adjudicator's Contract, he acts impartially, and if he resigns or is unable to act the parties jointly appoint a new Adjudicator. If the parties have not appointed an Adjudicator, either party may ask the nominating body to choose an Adjudicator within four days of the request.

A party may first give a notice of adjudication to the other party with a brief description of the dispute, details of where and when the dispute has arisen, and the nature of the redress which is sought.

The Adjudicator may be named in the contract in which case the party sends a copy of the notice to the Adjudicator. The Adjudicator must confirm within three days of receipt of the notice that he able to decide the dispute, or if he is unable to decide the dispute.

Within seven days of the issue of the notice of adjudication, the party:

- refers the dispute to the Adjudicator;
- provides the Adjudicator with the information on which he relies together with supporting information;
- provides a copy of the information and supporting documents to the other party.

Again, a dispute under a subcontract, which is also a dispute under the ECC, may be referred to the Adjudicator at the same time and the Adjudicator can decide the two disputes together.

The Adjudicator may review and revise any action or inaction of the Project Manager, take the initiative in ascertaining the facts, and the law relating to the dispute, instruct a party to provide further information and instruct a party to take any other action which he considers necessary to reach his decision.

All communications between a party and the Adjudicator must be communicated to the other party at the same time.

The Adjudicator decides the dispute and notifies the parties and the Project Manager within four weeks of the end of the period for providing information. This four-week period may be extended by joint agreement between the parties. Until this decision has been communicated, the parties proceed as if the matter in dispute was not disputed.

The Adjudicator's decision is binding on the parties unless and until revised by a tribunal and is enforceable as a contractual obligation on the parties. The Adjudicator's decision is final and binding if neither party has notified the Adjudicator that they are dissatisfied with an Adjudicator's decision within the time stated in the contract, and intends to refer the matter to the tribunal.

The Adjudicator may, within two weeks of giving his decision to the parties, correct any clerical mistake or ambiguity.

Review by the tribunal

A dispute cannot be referred to the tribunal unless it has first been referred to the Adjudicator.

Again, the tribunal may be named by the Employer within Contract Data Part 1, normally litigation or arbitration. If arbitration is chosen the Employer must also state in Contract Data Part 1, the procedure, the place where the arbitration is to be held, and the person or organisation who will choose an Arbitrator if the parties cannot agree a choice, or if the named procedure does not state who selects the Arbitrator.

A party can, following the adjudication, notify the other party within four weeks of the Adjudicator's decision that he is dissatisfied. This is a time-barred right, as the dissatisfied party cannot refer the dispute to the tribunal unless it is notified within four weeks of the Adjudicator's decision, failure to do so will make the Adjudicator's decision final and binding.

Referral to the tribunal under Option W2 excludes the provision where the Adjudicator has not notified his decision within the time provided by the contract as it is governed by statutory requirements.

The tribunal settles the dispute and has the power to reconsider any decision of the Adjudicator and review and revise any action or inaction of the Project Manager or Supervisor.

It is important to note that the tribunal is not a direct appeal against the Adjudicator's decision, the parties have the opportunity to present further information or evidence that was not originally presented to the Adjudicator, and also the Adjudicator cannot be called as a witness.

10.14 NEC3 Adjudicator's Contract

Introduction

The first edition of the NEC Adjudicator's Contract was published in 1994, and was written for the appointment of an Adjudicator for any contract under the NEC family of standard contracts. The second edition, published in 1998, contained some changes including the need to harmonise with the NEC standard contracts and further editions which had been issued since 1994. The third edition harmonises the contract with the NEC3 family of contracts using either Option W1 or W2.

The NEC3 Adjudicator's Contract can also be used with contracts other than NEC3, though dependent on the contract and the applicable law, some amendment may be necessary.

In the UK, the Housing Grants, Construction and Regeneration Act 1996 has made adjudication mandatory as a means of resolving disputes in construction contracts which fall under the Act. Parties to a contract which does not provide for adjudication as required by the Act have a right to adjudication under the Scheme for Construction Contracts. Schemes which are substantially similar have been published for England and Wales, Scotland and Northern Ireland.

The contract consists of five pages plus the index.

The agreement is between the two disputing parties and the Adjudicator.

The contract contains the following four sections.

(i) General

This covers the obligation upon the Adjudicator to act impartially, and to notify the parties as soon as he becomes aware of any matter which could present a conflict of interest.

There is a definition of expenses, which includes printing costs, postage, travel, accommodation and the cost of any assistance with the adjudication. All communications must be in a form which can be read, copied and recorded. This is a requirement of all NEC contracts, but also of the adjudication process itself, with both parties and the Adjudicator having to be copied into any communication between the parties.

(ii) Adjudication

The Adjudicator cannot decide any dispute that is the same or substantially the same as his predecessor decided. He must make his decision and notify the parties in accordance with the contract, and in reaching his decisions he can obtain assistance from others, but must advise the parties before doing so, and also provide the parties with a copy of anything produced by the assisting party, so that the parties can be invited to comment. The Adjudicator's decision is to remain confidential between the parties, and following his decision the Adjudicator retains documents provided to him for the period of retention, which is stated in the Contract Data.

(iii) Payment

The Adjudicator's hourly fee is stated in the Contract Data. There is provision for an advance payment to be made by the referring party to the Adjudicator if stated in the Contract Data.

Following his decision, the Adjudicator submits an invoice to each party for their share of the amount due, including expenses. The parties pay the amount due within three weeks of receiving the Adjudicator's invoice. Interest is paid at the rate stated in the Contract Data on any late payment. The parties are jointly and severally liable to pay the amount due to the Adjudicator, therefore if one party fails to pay, the other must pay including any interest on the overdue amount, and recover the amount from the other party.

(iv) Termination

The parties may, by agreement, terminate the Adjudicator's agreement. Also, the Adjudicator may terminate the agreement if there is a conflict of interest, he is unable to decide a dispute, an advance payment has not been made, or he has not been paid an amount due within five weeks of the date by which the payment should have been made. The Adjudicator's appointment terminates on the date stated in the Contract Data.

10.15 The tribunal

Whilst Contract Data Part 2 requires the Employer to enter the form of the tribunal, it does not say what choices are available. Other contracts refer disputes to expert determination or Dispute Review Boards, but under the ECC the Employer normally selects arbitration or litigation. Within the Contract Data the Employer is required to state the arbitration procedure, the place where the arbitration is to be held, and who will choose an Arbitrator.

There are a number of differences between the two but essentially arbitration is hearing the dispute in private and litigation is hearing the dispute in a public court.

It must be stressed that the role and authority of the tribunal is not to review and appeal against the previous decision of an Adjudicator, but to resolve the dispute independent of the adjudication; therefore it is quite in order for parties to introduce new evidence or materials to the tribunal.

10.16 Arbitration

Any dispute between two or more parties can be resolved through the courts, a process known as litigation or legal proceedings.

However, litigation has many disadvantages, not least the cost of a full court hearing, but also in many cases a number of years of waiting time before the matter actually reaches the courts, so in recent times arbitration has become increasingly used as a simpler, more convenient method of dispute resolution.

Arbitration is the hearing of a dispute by a third party, who is often not a lawyer, though that term has a very wide meaning, but is an expert in the field in which the dispute is based, and can give a decision based on opinion, which is then legally binding on both parties.

Arbitration is not a new concept; in fact it has been in existence for almost as long as the law itself, the first official recognition in the UK being the Arbitration Act 1697, which largely governed disputes about the sale of livestock in cattle markets. With the continuing growth of freight transport by ship, and the advent of railways and other forms of transport, a large number of arbitration cases ensued, and as a result Parliament passed the Arbitration Act 1889.

Various amendments were introduced over the following years with the current legislation being the Arbitration Act 1950, as amended by the Arbitration Acts of 1975 and 1979. The Arbitration Act 1996 is the current legislation related to arbitration in England and Wales.

One does not issue writs in arbitration as one would with litigation; *both* parties agree to enter into arbitration as a pre-agreed provision within a contract and thus be bound by the decision of the Arbitrator.

Arbitrators are appointed by one of three methods:

1. As a result of a pre-agreed decision between the parties, normally through a provision within a contract to go to arbitration in the event of a dispute. In this case the Arbitrator himself may be named or subject to nomination by a professional body such as the Chartered Institute of Arbitrators.
2. By statute. Legislation in many countries includes the provision for disputes to be settled by arbitration.
3. By order of a court.

Primary arbitration bodies in the UK:

*   the Chartered Institute of Arbitrators;
*   the London Court of International Arbitration;
*   the London Maritime Arbitrators Association.

The Arbitration Act is fairly brief and sets out procedures in the absence of any agreement to the contrary between the parties.

As arbitration is a more flexible arrangement than litigation, whatever procedure the Arbitrator and both parties agree to is sufficient in an arbitration case.

Whilst most forms of contract provide for resolution of disputes by arbitration, it must be stressed that it should only be seen as a last resort when all attempts at negotiation and other resolution methods have failed.

It may be surprising to hear that arbitration is common as a settlement procedure for disputes under trade agreements, maritime and insurance, consumer matters such as package holidays, construction industry disputes and property valuations.

The contracting party that initiated that the dispute be referred to arbitration is referred to as the 'Claimant', whilst the other party is referred to as the 'Respondent'. Should there be a Joint Agreement to go to arbitration, the Arbitrator will decide who is the Claimant, and who is the Respondent.

It is important that these titles are clarified as the Arbitration Rules repeatedly refer to them.

There are several advantages in using arbitration rather than litigation.

(i) Privacy

This is a distinct advantage, as in litigation, private and often personal matters are debated in full view of the public in open court, and often published in the press and technical magazines, whereas arbitration is carried out in private sessions, usually at the Arbitrator's office or at a pre-arranged venue.

(ii) Convenience

Arbitration is carried out at a time and place to suit the parties, for example evenings and weekends, not when the court is available for session.

(iii) Speed

The dispute is settled efficiently without the normal delays and procedural matters involved in often fairly lengthy and formal litigation.

(iv) Simplicity

There are few technical procedures as in the courts, the parties set their own procedures within a set of ground rules.

(v) Expertise

The Arbitrator will be an expert on the matter in dispute, for example a Structural Engineer where the dispute is about structural design issues, whereas judges and other legal professional are experts only on principles of law. The dispute will often not hinge on a legal problem; for example, in matters of design who better than a Designer to resolve it?

(vi) Thoroughness

Because the Arbitrator is a skilled technical person, the dispute can often be dealt with more thoroughly and matters more rigorously aired than in a confined and formal court atmosphere.

(vii) Expense

It can be readily seen that arbitration is generally a less expensive option than litigation, though it must not be seen as a cheap way to resolve a dispute. In some complex cases arbitration can be even more expensive than litigation.

It is worth mentioning at this point that, in many countries, legal aid, which can be available for litigation, is not available for arbitration.

The main disadvantages of using arbitration rather than litigation are as follows.

(i) Lack of legal knowledge

A dispute may hinge on a difficult point of law, which the Arbitrator has insufficient legal knowledge to resolve, as he is an expert in the field relating to the dispute, not in the law itself. He is, however, entitled to seek legal advice on a point of law, but not on the dispute itself, and also the disputing parties can refer a question of law to the High Court.

NB The arbitration award itself cannot be referred to the courts unless *both* parties consent, or the courts are referred to on a point of law, the resolution of which could substantially overturn an Arbitrator's decision.

Further, Arbitrators cannot be sued for negligence as they do not hold themselves up as experts in the sense of being advisory bodies and therefore liable for giving wrongful advice.

(ii) Precedents

Arbitration is not subject to rulings in previous cases as with the courts, each case being judged on its own merits on the day, and also to some extent subject to the personality of the Arbitrator; therefore there is no guide as to how successful an action could be.

It is prudent to always seek professional advice before resorting to arbitration as whilst it is simpler than court action the Arbitrator's decision is final and binding, therefore a party must be confident that they have a strong case.

(iii) Decision making

It has been said that as the Arbitrator, being a technical person and possibly having been involved in a similar dispute himself, tends to be fairer than judges in litigation; Arbitrators also tend to sympathise with both parties and can split the difference between the parties when making their awards as they have possibly been in a similar position themselves, whereas judges are inclined to take colder decisions and award wholly in favour of one of the parties. It has to be said that this is not a commonly held view!

10.17 Reference to arbitration

Although arbitration is a different process to adjudication, the same initial principles apply before arbitration can commence in that there must be a dispute, there must be an agreement to arbitrate, which will normally be provided for within the contract, and a party must submit a notice to initiate the process.

Proceedings begin by either party sending a written notice to the other stating that they wish to refer the dispute to arbitration, the form and process may be dictated by the arbitration procedure stated by the Employer within Contract Data Part 1.

The parties may agree on a person to act as Arbitrator, or a professional body will select and appoint the Arbitrator. In some cases, where the parties cannot agree a choice and there is no provision for a third party to do it, a court may select an Arbitrator. There is normally only one Arbitrator, but in some cases three, consisting of two Arbitrators and a third who acts as an umpire. Any correspondence flowing between the two parties as from the date of appointment must then also be copied to the Arbitrator(s).

The next step is that the Arbitrator writes to the parties accepting the appointment, a Preliminary Meeting is held between the parties at a time and venue to be directed by the Arbitrator to decide the form of the hearing which may be on a documents-only basis or an oral hearing. Documents only tends to be cleaner and quicker as there is less opportunity for the parties to debate the issue.

10.18 The role of the Arbitrator

The role of the Arbitrator is only to examine the evidence before him enabling him to resolve the dispute, and nothing more, he cannot look into other matters not in dispute. He can conduct the arbitration as he wishes, though he must gear the procedures in a way that is not detrimental to either party.

He must also have a regard for the rules of evidence. For example:

- Hearsay evidence is not permissible.
- He is not permitted to use evidence from someone not qualified to give such evidence.
- The Arbitrator is not allowed to take into account any evidence he has discovered himself, which was not provided by either party.
- He is not allowed to use any privileged evidence, for example that which is marked 'Without Prejudice'.
- In addition, the rules of discovery apply where either party is entitled to have access to the file or documents of the other provided that it relates to matters at issue.

10.19 Inspection by the Arbitrator

The Arbitrator is entitled to inspect any property, work or Materials at any premises, and may invite the Claimant and/or the Respondent to accompany him purely for the purpose of identifying the work or Materials.

No one else is allowed to accompany the Arbitrator unless he specifically invites him.

10.20 The award

The award is binding upon both parties and must be served in the appropriate manner with full headings, summary of issues in dispute, outline of events leading to the dispute, his decision based on the hearing and finally the monetary award.

There is very limited scope for appeal against an Arbitrator's award. Usually, appeals can only be based on a claim that the Arbitrator was wrong on a point of law, or there was serious and proven irregularity.

Arbitration awards can also be enforced in court if necessary.

10.21 Arbitrator's fees and expenses

Both parties will normally incur costs, though it is the Arbitrator's decision as to how the costs are apportioned. Both parties are initially liable to the Arbitrator for payment of his fees and expenses, but the Arbitrator will direct the parties as to the proportion they are liable for once the decision is reached.

Should the parties agree on a settlement of the disput20e at any time during the arbitration, they will jointly be liable for the Arbitrator's costs.

10.22 Litigation

It is not proposed to go into any detail about the litigation process from initiation to judgement by a court, first as the contract does not give any details of the process, but also because the proceedings themselves will depend on the law of the contract, which in turn is usually dependent on the location of the project.

However, the process will normally involve the Claimant's representative, usually a Solicitor, issuing a claim form or writ to the appropriate court; papers are then served on the Defendant. The Defendant then replies to the service by either admitting or denying the claim against him; if he denies it, he replies by sending his defence which may also be accompanied by any counter-claim. The case is then allocated to a court. There then follows a period of accumulation, collation and presentation of evidence, and hearings, again the process, the submission and the form of the hearing being dependent upon the law of the contract, until the court passes judgement.

11 Tenders

11.1 Deciding the procurement strategy

Deciding the procurement strategy is the foundation stone of a successful outcome to a contract, but this process is very often understated or almost overlooked
as Employers and their advisers, in their urgency to get projects out to tender and
commenced on Site, tend to select the procurement strategy most familiar to
them, rather than to spend time properly considering the project in terms of the
Employer's resources and his objectives in terms of time, price and quality, with
the additional factor of risk having to also be considered and evaluated. It is rarely
achievable for all three to be fully satisfied; it is usually a compromise.

The initial briefing with the Employer is the fundamental first step in establishing what is required, when it is required, and what the Employer is proposing,
and/or can afford to pay. Through a process of interviewing key members of the
Employer's team and reviewing his business plans, specific objectives, options and
constraints, a procurement strategy is developed which forms the basis for future
assessments and recommendations.

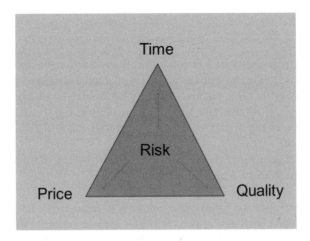

Figure 11.1 The time, price, quality and risk triangle

Whilst it may sound obvious, it is vital that the Employer understands what he wants and can also communicate it clearly. He may not know what he wants, but also he can often be swayed by consultants or other third parties into something he does not want.

In broad terms, the procurement strategy should consider time, price and quality and the balance between the three as follows:

(i) Time

Most Employers want their projects to be completed as quickly as possible. This is particularly relevant in sectors such as retail, where the Employer needs to open a shop as early as possible in order to sell goods. Also, an Employer may need to tie in Completion of a project with transfer of staff, students or patients from older buildings. And some projects have absolute deadlines, eg international sporting events.

If the Employer sees time as being the most important factor, he may have to pay more money, and quality may have to be lowered in order to achieve that. The procurement method will have to consider timing through programme requirements, but also through provisions such as sectional Completion and delay damages.

(ii) Price

All Employers will have some form of budget, so this will always be a major consideration. The budget may be fixed, representing all that the Employer has available through his own funds or borrowings in order to fund the project, or may have some degree of flexibility. The form of procurement, ranging from lump sum through to cost reimbursable, will need to address the budget and any required flexibility.

(iii) Quality

Finally, quality will always be a major factor, though if the above two are more important, many Employers will expect to lower standards either by not having what they want, or by value engineering, which considers similar but more economical solutions. It is important that the design lowers the standards by agreement, not that a 'cheap' Contractor is employed who will unilaterally lower the standards himself!

More detailed questions can then be asked and answered to finally complete the procurement strategy, which include the following:

- How important to the Employer is certainty of price, particularly if changes are expected? Lump sum options will give greater certainty of price, cost-reimbursable options will give less certainty of price, but greater flexibility.
- What resources and expertise does the Employer have? Cost reimbursable

and target contracts require resources to check invoices, time sheets, etc as opposed to lump sum or remeasurable contracts.

- Which party is to be responsible for design and/or which party has the necessary design expertise? Design can be assigned to the Contractor under any of the Main Options, though it is not recommended that bills of quantities are used for design and build.

- How important to the Employer is an early start and/or early Completion? Early commencement and early Completion may lead the Employer towards a cost reimbursable solution as lump sum and remeasurement procurement options require a considerable time to prepare the tender documentation for pricing.

- How clearly defined is the Scope of Works, and what form does it take, for example is it in the form of detailed drawings and specification, or performance-based criteria? Again, the contracts can be used with detailed drawings or performance-based criteria, where the Contractor carries out the design.

- What is the likelihood of change to those defined requirements? Many Employers prefer a remeasurement option where there is a likelihood of many changes. However, the ECC has a dynamic change management system (compensation events) which allows changes to be priced speedily.

- What is the Employer's attitude towards risk? Who is best placed to manage the inherent risks within the project? Lump sum contracts assign greatest risk to the Contractor, cost reimbursable contracts assign greater risk to the Employer.

  It is a poorly acknowledged fact that the Employer always pays for risk, regardless of whether it is carried within the contract by the Employer or Contractor. If it is the Contractor's risk he is deemed to have priced and programmed for it; if it is an Employer's risk he will pay for it if it occurs.

The procurement strategy then leads to a contract strategy which considers what is available in terms of contract to enable the Employer to achieve the objectives drawn from the procurement strategy.

Within the ECC, appropriate decisions can then be made through the Main and Secondary Options, and also whether the Contractor is to design all, or portions of the work.

Whilst many traditional contracts are based on bills of quantities with the Employer designing the project, and detailed bills of quantities being sent to each tendering Contractor for pricing, in the past 20 years there has been a movement away from the use of traditional bills of quantities and towards Contractor-designed projects with payment mechanisms such as milestone payments and activity schedules, with payment based on progress achieved, rather than quantity of work done.

11.2 Contractor design

This increasing growth of Contractor design has given rise to the increasing use of various design and build contracts, performance specifications and design, build, finance and operate projects.

There are a number of reasons for allocating some or most of the design to the Contractor:

- The design and construction periods can overlap, leading to faster delivery of the project.
- The Contractor can utilise his experience and preferred methods of construction to build rationally designed projects which minimise costs and programme durations.
- The temporary Works/permanent Works interface and influence of design is rationalised as both remain with the Contractor and therefore the permanent Works should be more 'buildable'.
- The traditional design/construction interface and the risks associated with it are transferred to the Contractor.
- The management of the design risk by the Contractor can result in greater certainty of the time, cost and performance and project objectives being met.

There has also been an increasing use in the past 15 years of target contracts which have been encouraged by the increasing use of partnering and collaborative working arrangements, and the associated more equitable sharing of risk.

Through its Main and Secondary Options, the ECC caters for all the procurement methods commonly used in construction contracts throughout the world.

11.3 Partnering arrangements

The past 15 years, particularly since the publication of the Latham and Egan reports, have seen the growth of partnering and framework agreements.

The US Construction Industry Institute has defined partnering as:

> A long-term commitment between two or more organizations for the purpose of achieving specific business objectives by maximizing the effectiveness of each participant's resources . . . the relationship is based upon trust, dedication to common goals and an understanding of each other's individual expectations and values.[1]

Partnering is a medium-to long-term relationship between contracting parties, whereby the Contractor, and in turn Consultants and various other parties, are not required to tender competitively on price for each project, but are awarded the work on the basis of their ability to deliver and normally by price negotiation.

The construction industry is known to be a high-risk business, and many projects can suffer unexpected cost and time overruns frequently resulting in disputes between the parties. The risks within a project are initially owned by the Employer, who may choose to adopt a 'risk transfer' approach where the risks are assigned through the contract to the Contractor who has the opportunity to price and programme for them, or a 'risk embrace' approach where the Employer retains the risks. In reality, most contracts are a combination of the two. The traditional approach to risk management is that of risk transfer, which is fine if the Scope of Works is clear and well defined; however, in recent years Employers have become more aware that they can achieve their objectives better by adopting a more 'old-fashioned' risk embrace culture.

Advantages of partnering agreements are:

- The procurement process from conception to commencement on Site, and subsequent Completion, can be significantly reduced in terms of time and resources.
- There is no re-tendering time and cost for future projects.
- Relationships can be developed based on trust.
- Contractors are appointed earlier and can contribute to the design and procurement process through ECI (Early Contractor Involvement).
- There tends to be greater cost certainty.
- Continuous improvement can be achieved by transferring learning from one project to the next.
- Better working relationships can be developed as the parties know each other.
- Prices can be more competitive and resources used more efficiently by continuous flows of work.

Disadvantages of partnering agreements are:

- Obligations and liabilities can become less clear in time as the parties can lapse into informal working patterns.
- Complacency can set in after some time as the Contractor has an assured flow of work.
- Employers are often dissatisfied that they are getting value for money in their projects.
- Continuing to award the work to a small number, or even one Contractor, prevents other equally good, or better, Contractors having the opportunity to carry out work.

All of these disadvantages can be overcome by maintaining a disciplined approach to communications between the parties and their rights and obligations, and also by continuous measurement of performance and deliverables.

11.4 Invitation to tender

As soon as it has been decided to select a Contractor by competitive tender rather than by negotiation, a shortlist of those considered suitable to be invited to tender should be compiled. Sometimes expressions of interest, interviews and presentations can be considered prior to drawing up the list.

It is critical that all the Contractors on the shortlist have the proven competence, resources, health and safety record, and financial stability to carry out the Works, so that when tenders are received and assessed, no doubt is in the

Employer's mind as to whether a preferred Contractor has the ability and resources to carry out the project. Employers should also review their shortlists periodically so as to exclude firms whose performance has been proven to be unsatisfactory, and also to allow the introduction of new Contractors.

The cost of preparing tenders is a significant element of the overheads of the Employer, the Contractors and in turn their Subcontractors and Suppliers, so lists of tenderers should be kept as short as practicable. Dependent on the size and type of project it is recommended that the shortlist should be from four to six names.

11.5 EU Procurement Directives

Whilst not wishing this book to be geography specific, it is worth some consideration of the detailed European Union (EU) Procurement Directives that apply to all procurement within the public sector, above minimum monetary thresholds which are reviewed on a regular basis, and subject to EU-wide principles of non-discrimination, equal treatment and transparency.

The purpose of the procurement rules is to open up the public procurement market and to ensure the free movement of supplies, services and works within the EU.

The rules are enforced by the Members States' Courts and the European Court of Justice and require that all public procurement must be based on value for money, defined as 'the optimum combination of whole-life cost and quality to meet the user's requirement', which should be achieved through competition, unless there are compelling reasons to the contrary.

In addition to the EU Member States, the benefits of the EU public procurement rules also apply to a number of other countries outside Europe because of an international agreement negotiated by the World Trade Organization (WTO) titled the 'Government Procurement Agreement' (GPA).

The EU Procurement Directives set out the legal framework for public procurement when public authorities and utilities seek to acquire supplies, services or works (eg civil engineering or building), procedures which must be followed before awarding a contract when its value exceeds set thresholds which are reviewed on a regular basis, unless the contract qualifies for a specific exclusion, for example on grounds of national security.

The Directives require contracting authorities to provide details of their proposed procurement in a prescribed format, which are then published in the *Official Journal of the European Union* (*OJEU*). All companies replying to an *OJEU* advertisement have an equal opportunity to express interest in being considered for tendering. As in all tendering exercises, Employers must ensure that those companies selected to tender receive exactly the same information on which to make their bid.

The notice may be issued electronically and this service may be provided through an intermediary organisation.

Generally, contracts covered by the Regulations must be the subject of a call for competition by publishing a Contract Notice in the *OJEU*.

Choice of procurement procedure

Four award procedures are provided for in the Regulations:

1. *the open procedure*, where all those interested may respond to the advertisement in the *OJEU* by tendering for the contract;
2. *the restricted procedure*, under which a selection is made of those who respond to the advertisement and only they are invited to submit a tender for the contract. This allows purchasers to avoid having to deal with an overwhelmingly large number of tenders;
3. *the competitive dialogue procedure*, under which, following an *OJEU* Contract Notice and a selection process, the authority then enters into dialogue with potential bidders, to develop one or more suitable solutions for its requirements and on which chosen bidders will be invited to tender; and
4. *the negotiated procedure*, under which a purchaser may select one or more potential bidders with whom to negotiate the terms of the contract. An advertisement in the *OJEU* is usually required but, in certain circumstances, described in the Regulations, no advertisement is required.

The *OJEU* Notice process has associated minimum timeframes for receipt of tenders or requests to participate, dependent on procurement procedure, from the date the Contract Notice is sent.

Stages in the procurement process

The Regulations set out criteria designed to ensure all Suppliers or Contractors established in countries covered by the rules are treated on equal terms, to avoid discrimination on the grounds of origin in a particular Member State.
     The criteria cover:

*   *specification stage* – how requirements must be specified;
*   *selection stage* – the rejection or selection of candidates;
*   *award stage* – the award of contract, either on the basis of 'lowest Price' or preferably the 'most economically advantageous tender' to the purchaser.

11.6 Preparing tender documents

The tender documents for an ECC contract will normally consist of:

*   *Invitation to Tender* (including instructions to tenderers on time and place of tender submission).
    There is no 'pro forma' version of 'Instructions to Tenderers' within the NEC contracts, but any invitation to tender will give a brief description of the work, who it is for, and how tenders may be submitted.

Other inclusions will cover such matters as:

- time, date and place for the delivery of tenders;
- what documents must be included with the tender, including a programme;
- the policy regarding alternative and/or non-compliant bids;
- arrangements for visiting the Site including contact details;
- rules on non-compliant bids;
- anti-collusion certificate.

- *Form of Tender.*
  There is no 'pro forma' Form of Tender within the NEC contracts, though the ECC Guidance Notes include a sample form. This is the tenderer's written offer to execute the work in accordance with the contract documents.
  The Form of Tender is normally in the form of a letter with blank spaces for tenderers to insert their name and other particulars, total tender Price, and other particulars of their offer. It is essential to have a standard Form of Tender and that all tenderers consistently use the same form, to make comparison of tenders easier.
- *The Pricing Document*, ie bills of quantities for an Option B or D contract, possibly a pro forma activity schedule for an Option A or C contract.
- *Works Information.*
  This will include descriptions of the Works including drawings, specifications and schedules, but also what would normally comprise 'Preliminaries' in a traditional contract (see 'Works Information' below).
- *Site Information.*
  This will include information about the Site and its surroundings including ground investigations, information from utilities companies (electricity, gas, water and telecommunications) regarding existing installations, and also environmental issues such a protected species, etc (see 'Site Information' below).
- *Contract Data Part 1* (completed by the Employer) (see 'Contract Data Part 1' below).
- *Contract Data Part 2* (blank pro forma to be completed by tendering Contractors) (see 'Contract Data Part 2' below).

In addition, there may be other information, dependent on the applicable legislation, for example in the UK, where the Construction (Design and Management) Regulations 2007 apply, Pre-Construction Information would be included within the tender documents.

11.7 Works Information

Works Information is defined by Clause 11.2(19) as 'information which either specifies and describes the Works or states any constraints on how the Contractor Provides the Works and is either in the documents which the Contract Data states it is in, or in an instruction given in accordance with this contract'. The Employer provides the Works Information and refers to it in Contract Data Part 1; if the Contractor provides Works Information, eg for his design, this is included in Contract Data Part 2.

The main documents within the Works Information are normally the drawings and specifications, but will also include:

Description of the Works:

- a statement describing the Scope of Works;
- schematic layouts, plan, elevation and section drawings, detailed working and/or production drawings (if relevant and available), etc;
- a statement of any constraints on how the Contractor Provides the Works, eg restrictions on access, sequencing or phasing of works, security issues.

Plant and Materials:

- Materials and workmanship specifications;
- requirements for delivery and storage;
- future provision of spares, maintenance requirements, etc.

Health and safety:

- specific health and safety requirements for the Site which the Contractor must comply with, particularly if the Site is within existing premises, including house and local safety rules, evacuation procedures, etc;
- any pre-construction information and health and safety plans for the project.

Financial records:

- details of the accounts and records to be kept by the Contractor.

Contractor's design:

- The default position is that all design will be carried out by the Employer. If any or all of the design is to be carried out by the Contractor, it should be included within the Works Information together with any performance requirements to be met within the Contractor's design. In addition, any warranty requirements and also any future novation requirements should be included within the Works Information.
- Any design acceptance procedures should be included including time scales for submission and acceptance of design.

Completion:

- Completion is defined under Clause 11.2(2), but the Works Information should also define any specific requirements which are required in order for Completion to have taken place for example, and requirement for:

- – 'as-built' drawings;
- – maintenance manuals;
- – training documentation;
- – test certificates;
- – statutory requirements and/or certification.

Services:

- details of other Contractors and Others who will be occupying the Working Areas during the contract period and any sharing requirements.

Subcontracting:

- lists of acceptable Subcontractors for specific tasks;
- statement of any work which should not be subcontracted;
- statement of any work which is required to be subcontracted.

Programme:

- Any information which the Contractor is required to include in the programme in addition to the information shown in Clause 31.2. Also, if the Employer requires the Contractor to produce a certain type of programme or to submit it using a certain brand of software, then this should be clearly detailed within the Works Information.

Tests:

- description of tests to be carried out by the Contractor, the Supervisor and Others, including those which must be done before Completion;
- specification of Materials, facilities and samples to be provided by the Contractor and the Employer for tests;
- specification of Plant and Materials which are to be inspected or tested before delivery to the Working Areas;
- definition of tests of Plant and Materials outside the Working Areas which have to be passed before marking by the Supervisor.

Title:

- statement of any Materials arising from excavation or demolition to which the Contractor will have title (Clause 73.2);
- requirements for Plant and Material to be marked (Clause 71.1).

Others:

- There are also certain specific requirements for statements to be made in the Works Information from certain Main and Secondary Options in the conditions of contract:

- X4.1, the form of any parent company guarantee;
- X14.2, the form of any advance payment bond;
- X13.1, the form of any performance bond.

The Works Information must be carefully drafted in order to define clearly what is expected of the Contractor in the performance of the contract and therefore included in the quoted tender amount and programme. If the contract does not cover all aspects of the work, either specifically or by implication, that aspect may be deemed to be excluded from the contract.

A comprehensive, all-embracing description should therefore be considered for the Scope of Works clause, which should be supplemented by specific detailed requirements. If reliance is placed solely on a very detailed scope description, an item may be missed from this detailed description and be the subject of later contention.

Where items of Equipment are to be fabricated or manufactured off Site by others, it is advisable that the contract sets out the corresponding obligations and liabilities of the respective parties, particularly if these are to form an integral or key part of the completed Works.

The Works Information describes clear boundaries for the work to be undertaken by the Contractor. It may also outline the Employer's objectives and explain why the work is being undertaken and how it is intended to be used. It says what is to be done (and maybe what is not included) in general terms, but not how to do it or the standards to be achieved. It explains the limits, where the work is to interface with other existing or proposed facilities. It may draw attention to any work or Materials to be provided by the Employer or others. It should also emphasise any unusual features of the work or contract, which tenderers might otherwise overlook.

This is the document that a tenderer can look to, to gain a broad understanding of the scale and complexity of the job and be able to judge its capacity to undertake it. It is written specifically for each contract. In some respects it is analogous to a shopping list. It should be comprehensive, but it should be made clear that it is not intended to include all the detail, which is contained in the drawings, specifications and schedules.

The Works Information will also include drawings which again should provide clear details of what the Contractor has to do. Clearly, tenderers must be given sufficient information to enable them to understand what is required and thus submit considered and accurate tenders.

It will also include the specification which is a written technical description of the standards and various criteria required for the work, and should complement the drawings. The Specification describes the character and quality of Materials and workmanship for work to be executed.

Again, it may lay down the order in which various portions of the work are to be executed. As far as possible, it should describe the outcomes required, rather than how to achieve them. It is customary to divide the work into discrete sections or trades (eg drainage, concrete, pavements, fencing, etc), with clauses written to cover the Materials to be used, the packaging, handling and storage of

Materials (only if necessary), the method of work to be used (only if necessary), installation criteria, the standards or tests to be satisfied, any specific requirements for Completion, etc.

The specification is an integral part of the design. This is often overlooked, with the result that inappropriate or outmoded specifications are selected, or replaced by a few brief notes on the drawings. The Designer should spend an appropriate amount of time specifying the quality of the work, as it is not possible to price, build, test or measure the work correctly unless this is done.

The Works Information describes what the Contractor has to do in terms of scope and standards, and in some cases must not do or include in order to Provide the Works. It may also include the order or sequence in which the Works are to be carried out. It also details where the work is to interface with other existing or proposed Contractors or facilities. It will also include any work or facilities or Materials to be provided by the Employer or others. It may also include any unusual features of the work, which tenderers might otherwise overlook, for example any planning constraints, etc.

This is the document from which a tenderer can gain a broad understanding of the scale and complexity of the job and his capacity to undertake it. It is written specifically for each contract.

The Works Information should be:

- *clear* – unambiguous;
- *concise* – not over-wordy;
- *complete* – have nothing missing.

In respect of the Contractor's design, a reason for not accepting the design is that it does not comply with either the Works Information or the applicable law. Again, if the Works Information was lacking and the Contractor provided a design that complied with it, and the applicable law, could the Contractor escape liability for the defective design? Secondary Option X15 states that 'the Contractor is not liable for Defects in the Works due to his design so far as he proves that he used reasonable skill and care to ensure that his design complied with the Works Information'. Again, this places a heavy burden on the Works Information.

11.8 Site Information

Site Information is defined in Clause 11.2(16) as 'information which describes the Site and its surroundings' and is identified in Contract Data Part 1.

Site Information may include:

- Ground investigations, borehole and trial pit records and test results. If the Employer has obtained such information, it should not be withheld from the tenderers, though the tenderers should be aware that the Site Information alone cannot be relied upon in terms of a possible compensation event (see Clause 60.2).

- Information about existing buildings, structures and Plant on or adjacent to the Site.
- Details of any previously demolished structures and the likelihood of any residual surface and subsurface Materials.
- Reports obtained by the Employer concerning the physical conditions within the Site or its surroundings. This may include mapping, hydrographical and hydrological information.
- All available information on the topography of the Site should be made accessible to tenderers, preferably by being shown on the drawings.
- Information about environmental issues, for example nesting birds and protected species.
- References to publicly available information.
- Information from utilities companies and historic records regarding Plant, pipes, cables and other services below the surface of the Site.

It is vital that care is taken to get the Site Information correct. Tenderers must be given sufficient information to enable them to understand what is required and thus to submit considered and well-priced tenders.

Under Clause 60.3, if there is an ambiguity or inconsistency within the Site Information, the Contractor is assumed to have taken into account the conditions most favourable to doing the work. Whilst many interpret that as the Contractor allowing the cheapest way of doing the work, it may be the easiest or quickest way.

In the event of the Contractor notifying a compensation event under Clause 60.1(12) for unforeseen physical conditions, he is assumed to have taken into account the Site Information, publicly available information referred to in the Site Information, information obtained from a visual inspection of the Site, and other information which an experienced Contractor could reasonably be expected to have or to obtain. Therefore the Contractor cannot rely solely on the Site Information in terms of how he prices and programmes the Works.

11.9 Provisional and Prime Cost Sums

Provisional and Prime Cost (PC) Sums are used in other contracts where there are elements of work which are not designed or cannot be sufficiently defined at the time of tender and therefore a sum of money is included by the Employer in the bill of quantities or other pricing document to cover the item. When the item is defined or able to be properly defined the Contractor is given the information which allows him to price it, the Provisional Sum is omitted and the Price included in its stead.

The problem with Provisional Sums is that they reduce the competition amongst tenderers as they are not priced at tender stage, and also if they are 'defined' the Contractor is deemed to have allowed time in his programme for them, if they are 'undefined' he has not.

If one considers General Rule 10 of the Standard Method of Measurement of Building Works Seventh Edition (SMM7), where work cannot be described and given in items in accordance with these rules it shall be given as a Provisional Sum and identified as for either 'defined' or 'undefined' work as appropriate.

Rule 10.3 of SMM7 requires that where the Provisional Sum is for defined work which is not completely designed, the following information shall be provided:

- the nature and construction of the work;
- a statement of how and where the work is fixed to the building and what other is to be fixed thereto;
- a quantity or quantities which indicate the scope and extent of the work;
- any specified limitations and the like identified in Section A35 of SMM7 (limitations imposed by the Employer, eg access, disposal of Materials found, working hours).

General Rule 10.4 of SMM7 states that 'where Provisional sums are given for defined work the Contractor will be deemed to have made due allowance in programming, planning and pricing preliminaries'.

General Rule 10.5 states that if the information specified in General Rule 10.3 is not available the Contract Bills should describe the provisional sum as 'undefined'.

General Rule 10.6 states that the Contractor is deemed to have made no such allowance for undefined work.

The NEC3 contracts do not provide for Provisional Sums, the principle being that:

- the Employer decides what he wants at tender stage so it can be designed and accurately described and the tendering Contractors can properly price and programme for them; or
- when the Employer decides what he wants, the Project Manager can give an instruction to the Contractor which changes the Works Information; it is a compensation event under Clause 60.1(1) and can be priced at the time. In that case, the cost of the work is held in the scheme rather than in the contract.

11.10 The Contract Data

The Contract Data provides the information required by the conditions of contract specific to a particular contract. Other contracts call this the 'appendices' or 'contract particulars'.

Part 1

Part 1 consists of data provided by the Employer, the sections of the Contract Data aligning with the sections of the core clauses.

Section 1: General

This section requires the Employer, or the party representing the Employer, to identify the selected Main and Secondary Options, a description or title for the Works, and the names of the Employer, the Project Manager, the Supervisor and the Adjudicator.

Whilst the Project Manager and the Supervisor must always be named individuals, many Employers choose to insert company names for these parties within the Contract Data and to separately identify the named individuals. It is also not uncommon for Employers to name directors or partners of their respective companies, then the authority of the Project Manager and Supervisor is delegated to the individuals under Clause 14.2.

The Works Information and Site Information are also identified within this section, though these are normally incorporated by reference to separate documents rather than listing drawing numbers and specification references. If this is the case, the separate documents must be clearly defined.

The boundaries of the Site are defined, normally by reference to a specific drawing or map.

As the NEC3 contracts are intended for use worldwide, the language and the law of the contract are also entered.

The period for reply is the period that the parties have to reply to submissions, proposals, notifications, etc, where there is no period specifically stated elsewhere within the contract, for example the period within which a party should reply to an early warning notice, or the Project Manager should reply to the Contractor submitting the particulars of his design or the name of a proposed Subcontractor. The period of reply would clearly not apply to the Contractor's submission of a programme for acceptance or the submission of a quotation for a compensation event.

With regard to dispute resolution, the Adjudicator Nominating Body is named within this section, normally a professional institution, and then the tribunal is named as either arbitration or legal proceedings.

Finally within this section, any matters to be included in the Risk Register will be identified.

Section 2: Contractor's main responsibilities

This section is not included within the Contract Data as no entries are required.

Section 3: Time

The starting date, access dates and the frequency of requirement for the Contractor to submit revised programmes are inserted within this section. Many Employers change the reference to 'one calendar month' rather than 'weeks', to align with monthly progress meetings or reporting requirements.

Section 4: Testing and Defects

The Defects Date is identified as a number of weeks after Completion of the whole of the Works. This identifies the period the Contractor is initially liable for correcting Defects.

The Defect Correction Period is also stated, this being the period within which the Contractor must correct each notified Defect, failing which the Project Manager assesses the cost to the Employer of having the Defect corrected by Others. The Contract Data provides for three entries to be inserted here if required, which then allows for different types of Defect to have different correction periods dependent on the urgency to have them corrected, for example:

- mechanical and electrical works – 24 hours;
- drainage – 48 hours;
- finishings – 7 days.

Clearly, any Defect that has a potential effect on health and safety should be corrected as soon as possible.

Section 5: Payment

Again, as the NEC3 contracts are intended for use worldwide, the currency of the contract is entered.

The assessment interval is also entered, which again is often expressed as 'one calendar month' rather than in 'weeks'. The Project Manager decides the first assessment date to suit the parties and following assessments are carried out within the assessment intervals.

Section 6: Compensation events

The only entries within this section are in respect of the weather, the place where weather is to be recorded (weather station, airport, etc) and the weather measurements to be recorded, the default measurements being cumulative monthly rainfall, the number of days with rainfall more than 5 mm, minimum air temperature less than 0 degrees Celsius, and snow lying at the designated time of day. The supplier of the weather measurements, the place where weather measurements are recorded, and where they are available from are also stated.

For some isolated Site locations where no recorded data may be available, assumed values may be inserted.

Section 7: Title

Again, this section is not included within the Contract Data as no entries are required.

Section 8: Risks and insurance

The minimum limit of indemnity for third party public liability and the Contractor's liability for his own employees is stated within this section.

Optional statements

Contract Data Part 1 contains a series of optional statements which are completed where appropriate.

First, if the tribunal has been identified in Section 1 as arbitration, the Employer must identify the arbitration procedure, the place where any arbitration is to be held, who will choose an Arbitrator if either the parties cannot agree a choice, or if the arbitration procedure does not state who selects an Arbitrator.

If the Employer has decided the Completion Date for the whole of the Works, the date is inserted in this section, alternatively the tendering Contractors may be required to insert a date in Contract Data Part 2.

When the Contractor completes the Works, the Project Manager certifies Completion and the Employer takes over the Works not later than two weeks after Completion, even though take-over may be before the Completion Date. If the Employer is not willing to take over the Works before the Completion Date, the statement to that effect should remain in the Contract Data; if not, it should be deleted. This statement is referred to in Clause 35.1.

The timing for submission of the Contractor's first programme is stated as a period of weeks from the Contract Date, the date when the contract came into existence.

If the Employer has identified work which is to meet a Key Date (Clause 11.2(9)), this is stated by the Employer as a condition to be met, and the Key Date.

If Option Y(UK)2 is not used, the period for payment is three weeks from the assessment date (Clause 51.2), though this can be amended in this section. Similarly, if Option Y(UK)2 is used, the final date for payment is 14 days after the date when the payment is due, and again this can be amended within the section.

If there are additional Employer's risks (Clause 80.1), these can be listed.

There are then optional statements regarding insurance.

First, if the Employer is to provide Plant and Materials there is provision for insurance of the Works to include any loss or damage of such Plant and Materials.

The Contractor provides the insurances stated within the Insurance Table, except any insurances which the Employer is to provide which are stated in Contract Data Part 1, which lists what the insurance is to cover, the amount of cover and the deductibles (otherwise known as 'excesses'), the amounts to be paid by the insuring party, often before the insurance company pays.

Dependent on which Main Option is selected, the Employer enters details for the relevant entry in the Contract Data.

- If Options B or D are used, the Employer states the method of measurement and any amendments.

- If Options C or D are used, the share range and Contractor's share percentage are stated.
- If Options C, D, E or F are used, the intervals within which the Contractor prepares forecasts of Defined Cost is defined. Also, the exchange rates to be used and the date on which they are published.

Finally, dependent on which Secondary Options are selected, the Employer enters details for the relevant entry in the Contract Data.

Part 2

Part 2 is completed by the Contractor and includes the names of key people and documents submitted with the tender, and also various percentages for Fee, people overheads, Working Areas overheads, design overheads, manufacture and fabrication overheads, and adjustment for listed Equipment, etc appropriate to the choice of Main Option. It is critical that these percentages are analysed as part of the tender assessment process.

11.12 Submission of tenders

There is a tendency to give Contractors very short periods of time in which to submit their tenders as the Employer wishes the work to be commenced as soon as possible; however, it is critical that tenderers are given sufficient time in which to properly consider what they have to price, and time- and resource-related issues, and to make any enquiries and obtain quotations prior to submitting their tenders.

The time and place for the delivery of tenders needs to be clearly stated. It is also critical that all tenderers are treated on an equal basis; therefore it should be made clear to all tenderers that late returns will be invalid and will not be considered. Late tenders should be returned to the tenderer unopened.

The tender documentation should also make clear the period over which a tender may be required to remain open for acceptance. This would normally be three months, giving the Employer and his advisers time to properly assess the tenders and make the appropriate award.

11.13 Opening tenders

Again, in order to maintain equality between the tenderers, tenders should not be opened until after the date and time for submission to avoid any suspicion of collusion. No tenderer should be allowed to alter his tender after the closing time.

After opening, the tenders and supporting documents should be evaluated. Each tender is confidential and no details of a tender should be released to any other tenderer.

11.14 Notification of tender results

It is important that all tendering Contractors be informed of the results of tendering as soon as possible after the tender submission date. This applies particularly to those whose tenders are not successful, or which are not being given further consideration, as it frees them to concentrate on opportunities with other Employers.

Generally, Employers' representatives have no authority to unilaterally accept tenders, as ultimately the contract is between the Employer and the Contractor, so it is normal to recommend a choice to the Employer and allow them the final decision and inform the tenderers.

The Employer's representative should submit a report, which lists the tenders received, comments on them and recommends a particular one for acceptance. The scope and detail of this report will vary according to the circumstances. The report should present a clear and logically reasoned case for acceptance of the recommended tender.

The following may be included in the report:

- a tabular statement of all the tenders received, showing the name of each tenderer, the tender sum, whether it was a valid submission, and any qualifying conditions;
- reference to any discussions that may have taken place with any tenderers regarding the tender;
- a concise summary of the findings following the examination and assessment of each tender, reasons for considering any tender invalid, any discussion of the programme and methods proposed, and reasons for considering any of these to be unsatisfactory;
- comments on any rates or Prices which appear exceptionally high or low, and forecasts of the possible effects on the contract Price in cases where there are large differences;
- a comparison of the tender of the recommended successful bidder with the original cost plan or budget;
- recommendation of the most acceptable tender.

REMEMBER – THE CHEAPEST PRICE DOES NOT
ALWAYS MEAN THE CHEAPEST COST!

As discussed earlier, by carrying out a detailed pre-tender analysis when tenders are received, the evaluation criteria should, in theory, be fairly straightforward as all the tenders should be capable of being accepted, so it will normally just include any further information gained from the bid itself.

11.15 Awarding the contract

There is no Standard Form of Agreement for completion and signature within the NEC3 contracts, though within the ECC Guidance Notes there is a sample form.

The drafters of the contract have stated that parties enter into contracts in many different ways, by executing Forms of Agreement, by exchange of letters, etc. Whilst this is probably true, it would have assisted the parties if there had been a standard Form of Agreement that they could use if they so wish.

The difficulty with not having a Form of Agreement within the contract is that the parties have to write their own; they also have to consider whether the contract is to be executed as a deed, and in that respect they must be very cautious that all the documents that are intended to form the contract are included in the Form of Agreement.

Note

1 Construction Industry Institute (CII) Partnering Task Force (1987).

Index

References in *italics* indicate figures